D0315372

Primary SPACE Project Research Team

Research Co-ordinating Group

Professor Paul Black (Co-director)
Jonathan Osborne

Centre for Educational Studies
King's College London
University of London
Cornwall House Annexe
Waterloo Road
London SE1 8TZ

Tel: 071 872 3094

Prof. Wynne Harlen (Co-director)
Terry Russell (Deputy Director)

Centre for Research in Primary
Science and Technology
Department of Education
University of Liverpool
126 Mount Pleasant
Liverpool L3 5SR

Tel: 051 794 3270

Project Researchers

Pamela Wadsworth (from 1989)
Bob Austin (from 1990)

Derek Bell (from 1989)
Ken Longden (from 1989)
Adrian Hughes (1989)
Linda McGuigan (from 1989)
Dorothy Watt (1986-89)

LEA Advisory Teachers

Maureen Smith (1986-89)
(ILEA)

Joan Boden
Karen Hartley
Kevin Cooney (1986-88)
(Knowsley)

Joyce Knaggs (1986-88)
Heather Scott (from 1989)
Ruth Morton (from 1989)
(Lancashire)

PRIMARY SPACE PROJECT
RESEARCH REPORT

January 1993

Rocks, Soil and Weather

by
TERRY RUSSELL, DEREK BELL,
KEN LONGDEN and LINDA McGUIGAN

LIVERPOOL UNIVERSITY PRESS

First published 1993 by
Liverpool University Press
PO Box 147, Liverpool L69 3BX

British Library Cataloguing in Publication Data
Data are available
ISBN 0 85323 497 3

Printed and bound by
Redwood Press Limited, Melksham, Wiltshire

CONTENTS

Page

ACKNOWLEDGEMENTS

The research reported here was made possible by the support of the Nuffield Foundation, the publishers CollinsEducational, and Lancashire, Knowsley and Liverpool Education Authorities.

For the word-processing, thanks are due to Jean Blackmore, Dolores Semple and Rosie Diver.

INTRODUCTION

This introduction provides an overview of the SPACE Project and its programme of research.

The Primary SPACE Project is a classroom-based research project which aims to establish

* *the ideas which primary school children have in particular science concept areas.*

* *the possibility of children modifying their ideas as the result of relevant experiences.*

The research is funded by the Nuffield Foundation and the publishers, CollinsEducational, and is being conducted at two centres, the Centre for Research in Primary Science and Technology, Department of Education, University of Liverpool and the Centre for Educational Studies, King's College, London. The joint directors are Professor Wynne Harlen and Professor Paul Black. The following local education authorities have been involved: Inner London Education Authority, Knowsley and Lancashire.

The Project is based on the view that children develop their ideas through the experiences they have. With this in mind, the Project has two main aims: firstly, to establish (through an elicitation phase) what specific ideas children have developed and what experiences might have led children to hold these views; and secondly, to see whether, within a normal classroom environment, it is possible to encourage a change in the ideas in a direction which will help children develop a more 'scientific' understanding of the topic (the intervention phase).

In the first phase of the Project from 1987 to 1989 eight concept areas were studied:

Electricity
Evaporation and condensation
Everyday changes in non-living materials
Forces and their effect on movement
Growth
Light
Living things' sensitivity to their environment
Sound

In the second phase of the Project during 1989 and 1990, a further ten concept areas were studied:

Rocks, Soils and Weather
Earth in space
Energy
Genetics and evolution
Human influences on the Earth
Processes of life
Seasonal changes
Types and uses of materials
Variety of life
Weather

Research Reports are usually based on each of these concept areas; occasionally where the areas are closely linked, they have been combined in a single report.

The Project has been run collaboratively between the University research teams, local education authorities and schools, with the participating teachers playing an active role in the development of the Project work.

Over the life-span of the Project a close relationship has been established between the University researchers and teachers, resulting in the development of techniques which advance both classroom practice and research. These methods provide opportunities, within the classroom, for children to express their ideas and to develop their thinking with the guidance of the teacher, and also help researchers towards a better understanding of children's thinking.

The Involvement of the Teachers

Schools and teachers were not selected for the Project on the basis of a particular background or expertise in primary science. In the majority of cases, two teachers per school were involved. This was advantageous in providing mutual support. Where possible, the Authority provided supply cover for the teachers so that they could attend Project sessions for preparation, training and discussion during the school day. Sessions were also held in the teachers' own time, after school.

The Project team aimed to have as much contact as possible with the teachers throughout the work to facilitate the provision of both training and support. The diversity of experience and differences in teaching style which the teachers brought with them to the Project meant that achieving a uniform style of presentation in all classrooms would not have been possible, or even desirable. Teachers were encouraged to incorporate the Project work into their existing classroom organisation so that both they and the children were as much at ease with the work as with any other classroom experience.

The Involvement of Children

The Project involved a cross-section of classes of children throughout the primary age range. A large component of the Project work was classroom-based, and all of the children in the participating classes were involved as far as possible. Small groups of children and individuals were selected for additional activities or interviews to facilitate more detailed discussion of their thinking.

The Structure of the Project

In the first phase of the Project, for each of the concept areas studied, a list of concepts was compiled to be used by researchers as the basis for the development of work in that area. These lists were drawn up from the standpoint of accepted scientific understanding and contained concepts which were considered to be a necessary part of a scientific understanding of each topic. The lists were not necessarily considered to be statements of the understanding which would be desirable in a child at age eleven, at the end of the Primary phase of schooling. The concept lists defined and outlined the area of interest for each of the studies; what ideas children were able to develop was a matter for empirical investigation.

In the second phase of the Project, the delineation of the concept area was informed by the National Curriculum for Science in England and Wales. The concept area was broken into a number of themes from which issues were selected for research. Themes sometimes contained a number of interlocking concepts; in other instances, they reflected only one underlying principle.

Most of the Project research work can be regarded as being organised into two major phases each followed by the collection of structured data about children's ideas. These phases called 'Exploration' and 'Intervention', are described in the following paragraphs and together with the data collection produce the following pattern for the research.

Phase 1a	*Exploration*
Phase 1b	*Pre-Intervention Elicitation*
Phase 2a	*Intervention*
Phase 2b	*Post-Intervention Elicitation*

The Phases of the Research

For the first eight concept areas, the above phases were preceded by an extensive pilot phase. Each phase, particularly the pilot work, was regarded as developmental; techniques and procedures were modified in the light of experience. The modifications involved a refinement of both the exposure materials and the techniques used to elicit ideas. This flexibility allowed the Project team to respond to unexpected situations and to incorporate useful developments into the programme.

Pilot Phase

There were three main aims of the pilot phase. They were, firstly to trial the techniques used to establish children's ideas, secondly, to establish the range of ideas held by primary school children, and thirdly, to familiarise the teachers with the classroom techniques being employed by the Project. This third aim was very important since teachers were being asked to operate in a manner which, to many of them, was very different from their usual style. By allowing teachers a 'practice run', their initial apprehensions were reduced, and the Project rationale became more familiar. In other words, teachers were being given the opportunity to incorporate Project techniques into their teaching, rather than having them imposed upon them.

Once teachers had become used to the SPACE way of working, a pilot phase was no longer essential and it was not always used when tackling the second group of concept areas. Moreover, teachers had become familiar with both research methodology and classroom techniques, having been involved in both of them. The pace of research could thus be quickened. Whereas pilot, exploration and intervention had extended over two or three terms, research in each concept area was now reduced to a single term.

In the Exploration phase children engaged with activities set up in the classroom for them to use, without any direct teaching. The activities were designed to ensure that a range of fairly common experiences (with which children might well be familiar from their everyday lives) was uniformly accessible to all children to provide a focus for their thoughts. In this way, the classroom activities were to help children articulate existing ideas rather than to provide them with novel experiences which would need to be interpreted.

Each of the topics studied raised some unique issues of technique and these distinctions led to the Exploration phase receiving differential emphasis. Topics in which the central concepts involved long-term, gradual changes, such as 'Growth', necessitated the incorporation of a lengthy exposure period in the study. A much shorter period of exposure, directly prior to elicitation was used with topics such as 'Light' and 'Electricity' which involve 'instant' changes.

During the Exploration phase teachers were encouraged to collect their children's ideas using informal classroom techniques. These techniques were:

i **Using Log-Books (free writing/drawing)**

Where the concept area involved long-term changes, it was suggested that children should make regular observations of the materials, with the frequency of these depending on the rate of change. The log-books could be pictorial or written, depending on the age of the children involved, and any entries could be supplemented by teacher comment if the children's thoughts needed explaining more fully. The main purposes of these log-books were to focus attention on the activities and to provide an informal record of the children's observations and ideas.

ii **Structured Writing/Annotated Drawing**

Writing or drawings produced in response to a particular question were extremely informative. Drawings and diagrams were particularly revealing when children added their own words to them. The annotation helped to clarify the ideas that a drawing represented.

Teachers also asked children to clarify their diagrams and themselves added explanatory notes and comments where necessary, after seeking clarification from children.
Teachers were encouraged to note down any comments which emerged during dialogue, rather than ask children to write them down themselves. It was felt that this technique would remove a pressure from children which might otherwise have inhibited the expression of their thoughts.

iii ***Completing a Picture***
Children were asked to add the relevant points to a picture. This technique ensured that children answered the questions posed by the Project team and reduced the possible effects of competence in drawing skills on ease of expression of ideas. The structured drawings provided valuable opportunities for teachers to talk to individual children and to build up a picture of each child's understanding.

iv ***Individual Discussion***
It was suggested that teachers use an open-ended questioning style with their children. The value of listening to what children said, and of respecting their responses, was emphasised as was the importance of clarifying the meaning of words children used. This style of questioning caused some teachers to be concerned that, by accepting any response whether right or wrong, they might implicitly be reinforcing incorrect ideas. The notion of ideas being acceptable and yet provisional until tested was at the heart of the Project. Where this philosophy was a novelty, some conflict was understandable.

In the Elicitation which followed Exploration, the Project team collected structured data through individual interviews and work with small groups. The individual interviews were held with a random, stratified sample of children to establish the frequencies of ideas held. The same sample of children was interviewed pre- and post-Intervention so that any shifts in ideas could be identified.

Intervention Phase

The Elicitation phase produced a wealth of different ideas from children, and produced some tentative insights into experiences which could have led to the genesis of some of these ideas. During the Intervention, teachers used this information as a starting point for classroom activities, or interventions, which were intended to lead to children extending their ideas. In schools where a significant level of teacher involvement was possible, teachers were provided with a general framework to guide their structuring of classroom activities appropriate to their class. Where opportunities for exposing teachers to Project techniques had been more limited, teachers were given a package of activities which had been developed by the Project team.

Both the framework and the Intervention activities were developed as a result of preliminary analysis of the Pre-Intervention Elicitation data. The Intervention strategies were:

(a) ***Encouraging children to test their ideas.***
It was felt that, if pupils were provided with the opportunity to test their ideas in a scientific way, they might find some of their ideas to be unsatisfying. This might encourage the children to develop their thinking in a way compatible with greater scientific competence.

(b) ***Encouraging children to develop more specific definitions for particular key words.***
Teachers asked children to make collections of objects which exemplified particular words, thus enabling children to define words in a relevant context, through using them.

(c) ***Encouraging children to generalise from one specific context to others through discussion.***
Many ideas which children held appeared to be context-specific. Teachers provided children with opportunities to share ideas and experiences so that they might be enabled to broaden the range of contexts in which their ideas applied.

(d) ***Finding ways to make imperceptible changes perceptible.***
Long-term, gradual changes in objects which could not readily be perceived were problematic for many children. Teachers endeavoured to find appropriate ways of making these changes perceptible. For example, the fact that a liquid could 'disappear' visually yet still be sensed by the sense of smell - as in the case of perfume - might make the concept of evaporation more accessible to children.

(e) ***Testing the 'right' idea alongside the children's own ideas.***
Children were given activities which involved solving a problem. To complete the activity, a scientific idea had to be applied correctly, thus challenging the child's notion. This confrontation might help children to develop a more scientific idea.

(f) ***Using secondary sources.***
In many cases, ideas were not testable by direct practical investigation. It was, however, possible for children's ideas to be turned into enquiries which could be directed at books or other secondary sources of information.

(g) ***Discussion with others.***
The exchange of ideas with others could encourage individuals to reconsider their own ideas. Teachers were encouraged to provide contexts in which children could share and compare their ideas.

In the Post-Intervention Elicitation phase the Project team collected a complementary set of data to that from the Pre-Intervention Elicitation by re-interviewing the same sample of children. The data were analysed to identify changes in ideas across the sample as a whole and also in individual children.
These phases of Project work form a coherent package which provides opportunities for children to explore and develop their scientific understanding as a part of classroom activity, and enables researchers to come nearer to establishing what conceptual development it is possible to encourage within the classroom and the most effective strategies for its encouragement.

The Implications of the Research

The SPACE Project has developed a programme which has raised many issues in addition to those of identifying and changing children's ideas in a classroom context. The question of teacher and pupil involvement in such work has become an important part of the Project, and the acknowledgement of the complex interactions inherent in the classroom has led to findings which report changes in teacher and pupil attitudes as well as in ideas. Consequently, the central core of activity, with its data collection to establish changes in ideas should be viewed as just one of the several kinds of change upon which the efficacy of the Project must be judged.

The following pages provide a detailed account of the development of the Rocks, Soil and Weather topic, the Project findings and the implications which they raise for science education.

The research reported in this and the companion research reports, as well as being of intrinsic interest, informed the writing and development with teachers of the Primary SPACE Project curriculum materials, published by HarperCollins Educational.

1. METHODOLOGY

This report draws together research, carried out by the SPACE Project, into children's ideas relating to concepts involved in the study of Earth Sciences. Two separate periods of data collection were undertaken. The first took place between May and July 1989 and focused on concepts related to the weather. The second research period was from May to July 1990 and concentrated on concepts associated with rocks and soils.

The samples and research programmes are described in the sections that follow, firstly for the Rocks and Soil research (section 1.1) then for the Weather study (section 1.2). Section 1.3 delineates the concept areas and foci of the research.

1.1 Rocks and Soil Research

a) Schools

Seven schools from Lancashire Local Education Authority participated in this research. They are situated in the region of Preston and Ormskirk and include schools in both urban and rural areas. Six of the schools cover the whole of the primary age range while the seventh school was a junior school covering Year 3 to Year 6. Thus, the sample included children from Reception (usual age five) to Year Six (usual age eleven).

Names of schools, Head Teachers and teachers are reported in Appendix 1.

b) Teachers

Eleven teachers, all with previous experience of SPACE research, were involved in this part of the programme. Thus, they were aware of the SPACE philosophy and techniques from their work on other topics within the Project.

A general principle was that, where possible, each school should have two participating teachers for the purpose of mutual support; in each of three schools, however, one teacher worked alone.

Teachers received further support through two whole-group meetings which were scheduled for two hours in an afternoon. At the first of these meetings the teachers, advisory teachers and researchers discussed and planned ways in which they could find out children's ideas. At the second, ways of helping children to develop their ideas were considered. Throughout the research period, visits to classes were made by the Project team, including an advisory teacher.

c) *Children*

All children in the eleven classes were involved in the Project work to some extent. A stratified random sample of children was selected for interview by members of the Project team at two stages during the work. Teachers were asked to assign each child in their class to an achievement band (high, medium, low) related to their overall school performance. The interview sample was then randomly selected from the class lists so that numbers were balanced by achievement band and gender. This same sample was used by some teachers where data collection necessitated one-to-one work between child and teacher. In most cases, however, teachers collected information from all children in their classes.

1.1.2 The Research Programme

Classroom work took place in two major phases. The first of these was called *Exploration.* It had two interconnected aspects: one was to expose children to the topic (Rocks and soil); the other was to elicit children's ideas in relation to that topic.

The first aspect, *exposure*, stems from the intention that the second, *elicitation*, should occur after children had had some opportunity to explore and handle different soils. By setting the context using real materials, it was hoped to collect a more considered view rather than catch children by surprise.

The exploration of children's ideas by the class teacher was followed by interviews with a random sample of the class. These interviews were conducted by researchers and the advisory teacher.

The second major phase was termed *intervention.* During the previous phase, teachers had been 'holding back' to encourage children to express their own ideas. In the intervention, teachers offered children experiences which gave them an opportunity to reflect on their ideas, test them out, discuss them and amend, reject or retain them.

This was followed by a limited post-intervention elicitation for the whole class coupled with interviews of the same sub-sample as before. The research was carried out between January and May 1990. The sequence of events is given below:

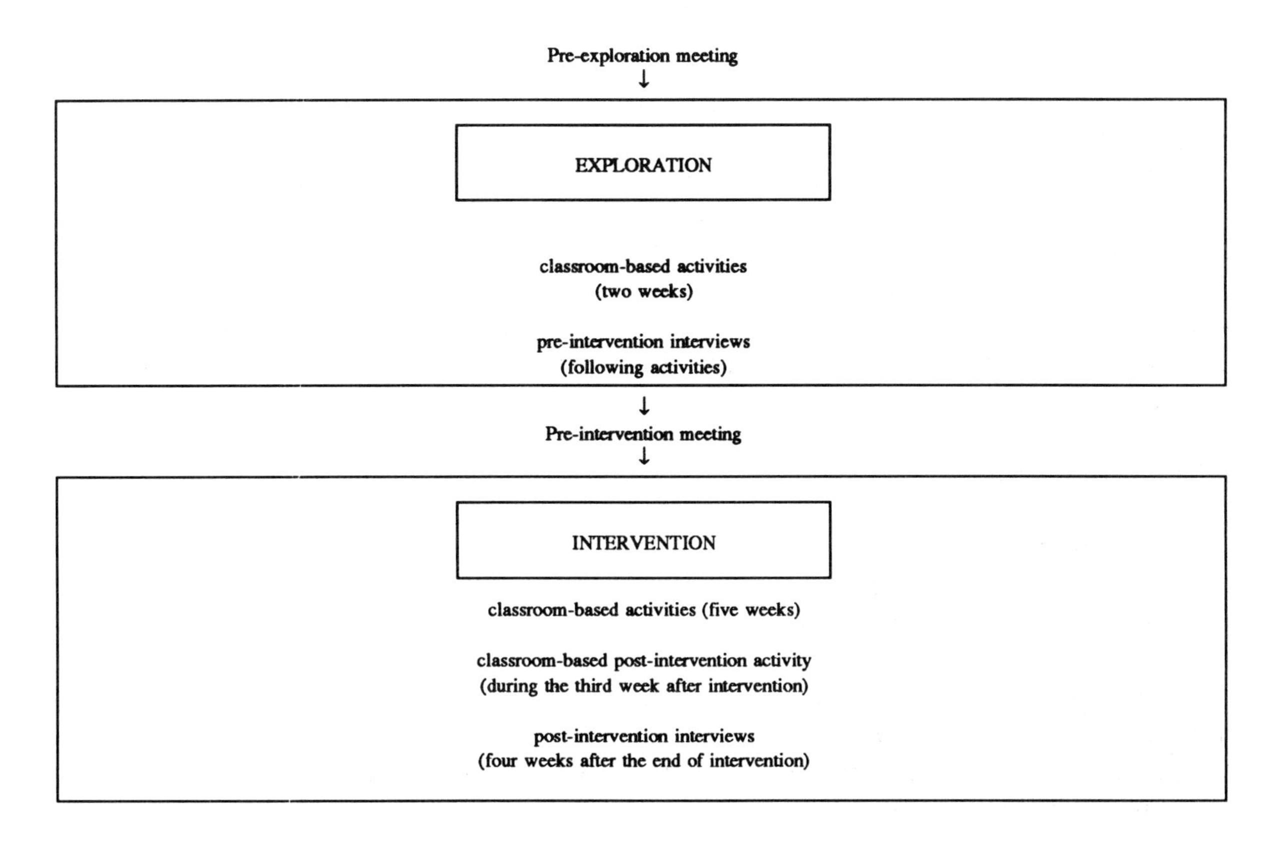

1.2 Weather Research

1.2.1 The Sample

a) Schools

Six schools from Knowsley Local Education Authority and one junior school from Liverpool Local Education Authority participated in this part of the research. All the schools are in urban areas which include Kirkby and Halewood on Merseyside. Five of the schools provide for the whole primary age range, while two junior schools covered Years 3 - 6. Thus, children from Reception (usual age five) to Year 6 (usual age eleven) were covered as part of the sample. Names of schools, Head Teachers, and teachers are given in Appendix II.

b) Teachers

All of the fourteen teachers who took part in this research had been involved in previous topics addressed by the Project. The teachers received similar support to that described for the Rocks and Soil Research, i.e. whole-group meetings which took place after school for about two hours, the pairing of teachers in schools and visits from researchers and advisory teachers.

c) *Children*

Teachers involved all children in their classes in the Project work. Although these children were not interviewed, teachers were asked to ensure that they collected all the appropriate data from all children. In practice, most of the teachers were able to obtain information from all members of their class.

1.2.2 *The Research Programme*

Two major phases of classroom work were carried out in a similar manner to that described for Rocks and Soil Research in Section 1.2 above. However, there was a major difference. This research programme did not include any individual interviews with children either pre- or post-intervention. The sequence of events is given below:

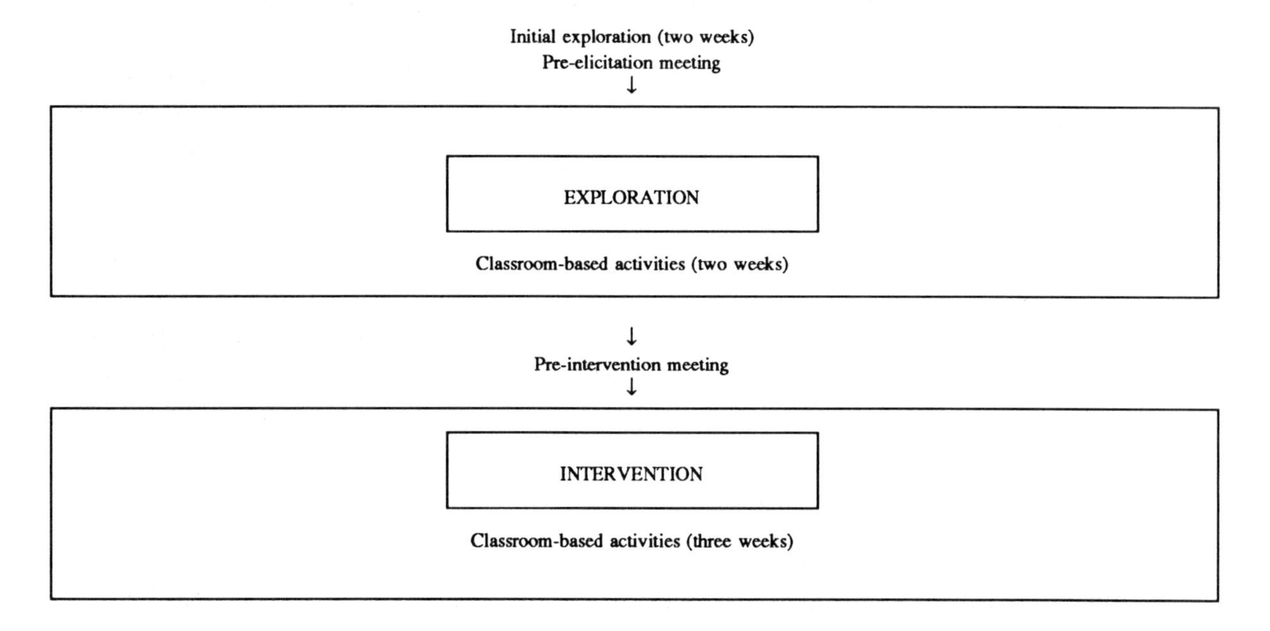

1.3 *Defining the Topic*

1.3.1 *Division into Themes*

In the early stages of the SPACE project a list of concepts was drawn up for each of the topics to be researched. This list was intended to delineate the boundaries within which the work would be carried out. With the advent of a National Curriculum for England and Wales, the document for Science provided a framework in which concepts for investigation could be identified (DES 1989).

The National Curriculum defines programmes of study (PoS) for different age groups of children. The programmes of study indicate the kinds of experiences to which all children of given age groups are expected to be exposed. For the primary age range of children, two programmes of study are relevant. That for Key Stage One (KS1) is applicable for children up to the school year in which they have their seventh birthday (infants) and that for Key Stage Two (KS2) being applicable for children up to the school year in which they have their eleventh birthday (juniors).

The National Curriculum also defines statements of attainment (SoA) applicable throughout the years of compulsory education. Statements of attainment are written for each of a series of ten levels. They indicate what a child must 'know', 'understand', or 'be able to' do, in order for attainment to be assigned to that level. The levels most likely to be achieved by primary aged children are levels one to five.

Each subject of the National Curriculum is divided into a number of Attainment Targets (ATs). Each target reflects a particular aspect of the subject which is deemed worthy of separate assessment. At the time when this research was planned and carried out, there were seventeen such targets designated for Science. The findings of this report pertain to AT9 'Earth and Atmosphere' as described in the original requirements of the Science National Curriculum document. Revision of the criteria in 1991 (DES 1991) reduced the number of Attainment Targets to four. The contents of this report refer to the new AT3 (Strand IV).

SPACE researchers divided the original AT9 into themes, each of which seemed to reflect a relatively discrete concept area within the topic of earth and atmosphere. Both programmes of study and statements of attainment for levels one to five were taken into account when making this sub-division. The outcome is shown as Figure 1.1. Programmes of study are preceded by 'KS1' or 'KS2' while statements of attainment are prefixed by a number denoting the level and a letter indicating its position within the level. Thus, statement 4c is the third statement at level four.

The new AT3 (Strand IV) has also been divided into themes using the revised programmes of study and statements of attainment. The outcome is presented in Figure 1.2.

Three themes were defined by SPACE researchers which can be identified in both Figure 1.1 and Figure 1.2; Theme A - Weather, Theme B - Earth, and Theme C - Linking Earth and Weather. Comparison of the two figures shows similarities in the concepts included, with Figure 1.1 (original AT9) providing a more detailed list of areas to be considered. It is particularly noticeable that Figure 1.2 (new AT3, strand IV) makes no reference to weather at KS1 and has no statements of attainment for weather below level 4. However, the themes outlined in Figure 1.1 were used to form the basis of the research and hence, this report.

(A)

Weather (Atmosphere)

KS1 Children should observe and record the changes in the weather and relate these to their everyday activities.

KS2 Children should have the opportunity to make regular, quantitative observations and keep records of the weather and the seasons of the year.

Pupils should:

1a know that there is a variety of weather conditions.

1b be able to describe changes in the weather.

2a know that there are patterns in the weather which are related to seasonal changes.

2b know that the weather has a powerful effect on people's lives.

2c be able to record the weather over a period of time, in words, drawings and charts or other forms of communication.

3b know that air is all around us.

3e be able to understand and interpret common meteorological symbols as used in the media.

4a be able to measure temperature, rainfall, wind speed and direction; be able to explain that wind is air in motion.

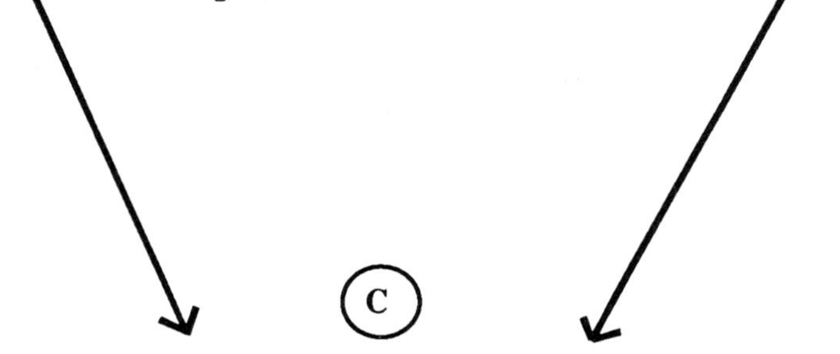

AT9 EARTH AND ATMOSPHERE

Pupils should develop their knowledge and understanding of the structure and main features of the Earth, the atmosphere and their changes over time.

(B)

Earth

KS1 Children should collect, and find differences and similarities in, natural materials found in their locality, including rocks and soil. They should compare samples with those represented or described at second hand.

KS2 Children should investigate natural materials (rocks, minerals, soils), should sort them according to simple criteria, and relate them to their uses and origins, using books and other sources. They should be aware of local distributions of some types of natural materials (sands, soils, rocks). They should observe, through urban or rural fieldwork, how weather affects natural materials (including plants) in their surroundings and how soil develops. They shoulld also consider the major geological events which change the surface of the Earth.

Pupils should:

2d be able to sort natural materials into broad groups according to observable features.

3d be able to give an account of an investigation of some natural material (rock or soil).

5a know that landscapes are formed by a number of agents including Earth movements, weathering, erosion and deposition, and that these act over different time scales.

5b be able to explain how earthquakes and volcanoes are associated with the formation of landforms.

Link to AT6 Types and uses of materials

(C)

Linking Earth and Weather (Atmosphere)

Pupils should:

3a be able to describe from their observations some of the effects of weathering on buildings and on the landscape.

3c understand how weathering of rocks leads to the formation of different types of soil.

4b know that climate determines the success of agriculture an understand the impact of occasional catastrophic events.
See also 5a (weathering, erosion)

5c be able to explain the water cycle.

Figure 1.1

AT3 Earth and Atmosphere

AT3 Strand IV

Weather (Atmosphere)

KS2 Pupils should have the opportunity to make regular quantitative observations and keep records of weather and the seasons of the year. This should lead to a consideration of the water cycle.

4d Know how measurements of temperature, rainfall, windspeed and direction describe the weather.

5d Understand the water cycle in terms of the physical processes involved.

Linking Earth and Weather (Atmosphere)

KS1 They should observe the effects of weathering in their locality.

KS2 They should observe through fieldwork, how weather affects their surroundings, how sediment is produced and how soil develops.

3c Understand some of the effects of weathering on buildings and on rocks.

4e Know that weathering, erosion and transport lead to the formation of sediments and different types of soils.

Earth

KS1 Pupils should observe and compare natural materials found in their locality, including rocks and soils.

KS2 Pupils should investigate natural materials (rocks, minerals, soils), sort them by simple criteria and relate them to their uses and origins. They should be aware of local distributions of some types of natural materials (sands, soils, rocks). They should consider the major geological events which change the surface of the Earth and the evidence for these changes.

1a Be able to describe the simple properties of familiar materials.

2a Be able to group materials according to observable features.

3b Know that some materials occur naturally while many are made from raw materials.

AT3 Strand 1 The properties, classification and structure of materials.

AT3 Strand 3 Chemical changes.

Figure 1.2

1.3.2 Issues Within the Themes

It was not possible to tackle all aspects of the themes rigorously and so particular issues were chosen to form the focus of the research. In particular, it is the first theme - Earth - which was most fully researched and which is reported in most detail.

1 Earth

This theme can be divided into three sub-themes which were useful in carrying out the research:

(a) Soil

What do children think soil is? What makes up soil? What substances do they consider to be soil? Where do children think soil comes from? How permanent do they consider soil to be? What changes might take place and what causes these changes?

(b) Rock

What do children understand by the term rock? Where do they think rocks are found? Do rocks extend underground? How do children think rocks are formed? How old do they think rocks are? Do they think rocks change?

(c) Earth's Structure

What do children think is under their feet? What is their understanding of the internal structure of the Earth? No direct attempts were made to elicit children's ideas about how particular landscape features or earth movements are brought about.

2. Weather

The following issues were selected from within this theme to form the focus of the research.

(a) Weather Records

What aspects of the weather do children notice? How do children make records of the weather? Do they use symbols? Are they aware of what the symbols mean? What ideas do they have about how people forecast the weather?

(b) Explaining Weather Changes

How do children explain changes in the weather? How do they explain changes in rainfall, temperature, cloud cover, wind speed and wind direction?

(c) The Effects of the Weather

What ideas do children have about how weather affects people, other animals and plants? Do children notice how weather affects the environment? To what extent are they aware of drought, erosion and weathering?

3. Links between Earth and Weather

This theme was not addressed separately but questions about the changes in rocks and soil and about the effects of weather on the environment were posed in both the Rocks and Soils and Weather themes. These were designed to gather children's ideas related to the issue of weathering in the environment.

2. *EXPLORATION*

2.1 Previous Research into Children's Ideas Related to Earth and Weather

Since the early work of Piaget (1929, 1930) much literature has been devoted to the ideas that children hold and bring to new learning situations. There now exists a significant body of information relating to children's understanding of concepts in physical and biological sciences, but rather less attention has been given to understanding of concepts related to earth sciences. With few exceptions, work investigating children's ideas in the various concept areas has concentrated on students of secondary age and above. Thus previous research into primary school children's ideas of rocks, soil and weather is very limited. The following account draws together much of the material relating to children's ideas across a range of concepts linked to rocks, soil and weather.

2.1.1 Weather

Piaget's extensive discussion (1929) of children's explanations of the various phenomena related to the weather and its causes forms the basis for any consideration of this topic. He interpreted responses to his questions about such things as clouds, rain, wind, snow and ice in terms of his stage theory. For example, when he questioned children (aged 5 to 11) about the formation of rain, Piaget suggested three stages of development. The first level of response explained the formation of rain as being made by God or man who 'throws out buckets of water' or 'from taps' (Piaget 1929 p 312). During the second stage the explanations suggest that the rain is produced from the products of human activity e.g. 'the smoke goes up and then makes clouds' and 'the smoke melts, it changes shape and then water comes' (Piaget 1929 p 314). The third stage of response explains rain in terms of 'air which is turned into water' and it rains 'because the clouds are wet. They are full of water.' (Piaget 1929 p 316). It is the **content** of children's responses within the 'clinical method' of interviewing which is of enduring interest.

Stepans and Kuehn (1985) in their study of children in USA in grades 2 and 5 found that the ideas expressed by these children reflected those identified by Piaget. Stepans and Kuehn went on to consider the type of instruction children received and suggested that children using a 'hands-on' approach had more accurate views about the phenomena associated with weather than children using textbooks'. However, this distinction was not so obvious with explanations of thunder and lightning which all children found difficult to explain.

Moyle (1980) studied children's ideas about weather as part of the Learning In Science Project (LISP), in New Zealand. The responses made by the children (aged 8 to 16) to questions about phenomena such as clouds, rain, wind were similar to the ideas reported by Piaget (1929). Children's responses to the question, 'What is weather?' were restricted to comments about the Sun, rain, wind and 'Bad' weather. In his discussion of the views of weather expressed, Moyle identifies ideas that are self-centred and human-centred, attribute human characteristics to features, and non-scientific cause and effect explanations. Children looked for causes in terms of concrete objects which can be detected by the senses. Thus aspects such as air pressure which are not directly detectable by the senses are less well understood.

Moyle (1980) also investigated children's perceptions of weather maps and found that most children (aged 8 to 16) had great difficulty in interpreting them. In particular he asked children to say what they thought the 'H's, 'L's and 'these lines' (isobars) meant. The 'H's were frequently associated with 'where it was going to be hot' and the 'L's with bad weather. Only four students (one aged 14 and three aged 16) related 'H's to high pressure and only one student (aged 16) explained 'L's represent areas of low pressure'. Very few students were able to offer any ideas as to what the lines (isobars) represented.

Bar (1989) carried out a detailed study of children's ideas of the water cycle and identified six phases related to children's development of ideas about the water cycle. The basis of each phase is related, in Bar's opinion, to the children's understanding of the underlying ideas of evaporation and conservation of materials. The phases outlined are:

a. Water and air are not conserved; during drying water disappears (ages 5 to 7). The associated ideas are:

  * Somebody - God - opens water reservoirs.
  * The clouds are kept above the sky; clouds are bags of water and rain falls when clouds open.

b. Water is conserved. Air is not conserved; during drying, water penetrates solid objects (ages 6 to 8).

  * Clouds enter the sea and collect water.

c. Water is conserved; Air is not conserved; during drying water penetrates solid objects; phase change occurs only during boiling (ages 6 to 9).

  * Clouds are made of water vapour created when the sea is heated by the sun to a high temperature or of water vapour originating in kettles.

d. Water and air are conserved; water evaporates into a container (ages 7 to 10).

* Clouds get scrambled and fall; clouds are shaken by the wind or sweat.

* Rain falls when the whole cloud turns into fog and humidity, or when it becomes hot or cold.

e. Water and air are conserved; water changes into vapour (ages 8 to 10).

* Clouds are made of water evaporated from the floor (the puddle); rain falls when the clouds become cold or heavy.

* Sea water as well as other sources evaporates to make clouds (ages 10 to 11).

f. Water and air are conserved; weight is attributed to air and water vapour and small drops of water (ages 11 to 15).

* The process of condensation and rainfall are distinguished.

Weather is a complex natural phenomenon which is difficult to understand and predict so it is not surprising that children are unsure of the mechanisms involved. Indeed, the lack of understanding of weather phenomena extends to undergraduate level and beyond as discussed by Nelson et al (1992). They identified a number of aspects of weather which are commonly described by undergraduates in terms that appear to conflict with the accepted scientific explanation.

2.1.2 Soil

Happs (1981, 1984) investigated children's ideas about soil as part of the LISP project. His study considered the views of students from the age of 11 to 17 plus two who were in higher education. Many children hold the view that soil has been around since the beginning of the Earth and is something in which plants grow as well as being a home for small creatures. Soil is considered to be static and unchanging, although it may be part of a cycle involving rocks and clay but no clear mechanisms are described. Children use different terms for soil. In particular, Happs refers to the fact that many of the children in his sample use the words 'dirt' and 'soil' synonymously.

When Happs (1981, 1984) asked children 'How far down do you think that soil would go?', if they dug down in their garden, the responses ranged from 6 inches to about 10 miles. When asked to indicate what they thought to be the depth of most soils in New Zealand the majority of students (aged 11 to 17) indicated depths from a few centimetres to a few metres. However, some students of all age groups suggested that soils were over several kilometres deep even extending to the earth's core, a depth of 3,000 kms from the Earth's surface.

2.1.3 Landforms and Earth Movements

Piaget (1929) questioned children about how they thought mountains were formed. The ideas expressed indicated that many children (aged 5 to 11) consider mountains to be made by man or God and that they were living. Some children did refer to that idea that the ground had risen up, thus linking mountain formation to earth movements. Happs (1982) took a slightly different approach in his study and asked students (aged 11 to 17) to consider two particular mountains found in New Zealand (Mt. Egmont and Mt. Cook).

The majority of children questioned by Happs talked of a mountain as being a large hill which had 'always' been there. In this case 'always' meant since the Earth was formed. Older students sometimes referred to the formation of mountains as being related to volcanoes and/or folding and faulting of the Earth's surface. Explanations of what is a volcano varied, the majority referring to eruptions in some form, but few recognised the example of Mt. Egmont as being a volcano. Occasionally, students hinted at a relationship with earthquakes. A small minority made reference to the idea of plate tectonics. The distribution of volcanoes was explained in terms of random occurrences - "They just pop up" - through association with climatic conditions (hot or cold regions, rarely both), to a firm link with mountains. Most of the students considered Mt. Cook to be a volcano although it is in fact the result of uplift of the land. Again, older students linked the formation of this mountain with earthquakes or plate tectonics.

Leather (1987) investigated children's ideas of earthquakes in more detail. Using a sample of children ranging in age from 11 to 17 he found the most common idea to be that earthquakes occur in hot countries and are caused by heat. Over half of the 11 year olds and a quarter of 16 year olds felt that earthquakes never happen in Britain. Ideas expressed by the younger children were still in evidence but given less frequently by the older children. In the same study, the children were asked about formation and occurrence of oil. At least a quarter of children in all age groups said that oil forms in caves under the seabed.

2.1.4 Rocks and Minerals

Happs (1982, reported in Happs 1985) carried out a study into children's understanding of rocks and minerals, and found that children (aged 11 to 17) expressed views which were quite different from the scientific explanations. In particular the word 'rock' was used by children in an everyday sense of 'a boulder-sized fragment of hard, dense and dark coloured rock'. The terms 'crystal', 'stone' and 'pebble' were also used in an everyday sense being loosely distinguished according to size and shape. Whilst the above terms were usually linked with that of 'rock', the word 'mineral' was rarely associated with rock of any description.

Happs (1985) in a refinement of his earlier investigation presents a detailed case study of a student's (aged 14) understanding of rocks and minerals. This reveals a number of interesting features associated with children's ideas about rocks and minerals. One such idea is the clear distinction many children make between rocks and stones. The former are large, dull, rough-looking objects while the latter are seen as small, smooth, round objects.

Evidence from the case study described by Happs (1985) indicates that students are unlikely to have an understanding of the terms 'sedimentary', 'igneous' and 'metamorphic' in relation to rocks and their formation. Grouping of rocks in terms of their origin was not suggested as a way of classifying rocks. Children used their own, everyday categories for sorting rocks into groups such as, 'shiny rocks' and 'ordinary rocks'.

2.2 Design of Exploration Experiences and Interview Questions

Design of exploration experiences and the interview questions was influenced by a number of factors:

a. the concepts, outlined in section 1.3, which were to form the focus of the research;

b. the previous research that has been summarised in 2.1 above;

c. the knowledge and expertise gleaned from previous SPACE research and the philosophy of the project.

This exploration phase of the research was considered to be exploratory in two senses. It was an exploration of children's thinking by the Project team. It was also an opportunity for children to start exploration of their own ideas. A sequence of activities was thus needed with the dual function of,

i *exposure* of children to the concept area of materials

ii *elicitation* of children's ideas.

This first aspect, exposure, reflects the Project's philosophy that elicitation should not be sprung on children. Through handling materials and gaining direct first hand experience children would have an opportunity to make clear their own thinking to themselves. Any view elicited would thus be more likely to be a considered one rather than one made up on the spur of the moment. Although it is true that such 'exposure' could cause reconsideration and change in ideas, this does not interfere with the Project's aim of obtaining a 'snapshot' of the kind of views that children hold about particular concepts.

Most of the teachers were familiar with the approach. Both they and research staff had become aware, from previous experience, of the kinds of activities that had proved successful in meeting the dual functions of 'exposure' and 'elicitation'. Such activities gave children a chance to think and also drew out their considered opinions. The successful activities had been based on certain techniques such as the use of open questions by the teacher in discussion with children. (Other techniques suitable for an exploration phase are described in the introduction to this report). More fundamentally, teachers were aware of the rationale for adopting those techniques. The techniques enabled them to establish children's ideas and it was those ideas which would serve as starting points in subsequent learning. This required teachers to adopt a role in which initially, they deliberately held back from guiding children's thinking. They were to help children to clarify their ideas rather than seeking to make them justify and reconsider them. The aim was to gather information on the issues outlined in 1.3 by combining responses to the classroom activities with those from the structured interviews.

2.3 Rocks and Soil Research

2.3.1 Exploration Experiences - Exposure and Elicitation

At the pre-elicitation meeting, the team of teachers, advisory teachers and researchers discussed and planned a series of exploration and elicitation activities. The number of suggestions made was, in practice, too many to be completed in the time available, so agreement was reached as to the ones which would be used. The guidelines where drawn up immediately after the meeting and sent out to all the teachers and advisory teachers. The full document is given in Appendix III.

The guidelines for the experiences were phrased in as non age-specific a way as possible, so that the common framework could be adapted by teachers of 5-year olds and by teachers of 11-year old children. In particular, teachers of younger children with limited writing skills had to rely more heavily on drawings and oral comments. These techniques were, of course, also available to the other teachers but, generally, they did not have to bear the same management demands of reporting data on behalf of the child as did the teachers of infants. It was thus necessary to provide a greater degree of classroom support to some teachers in order to free them for recording children's ideas.

Four basic experiences were carried out with the children but not necessarily in the order presented here.

1. Soil

   a. Looking at soil.

   Children were asked to bring samples of soil into school from their own gardens. Many of the teachers and some of the children obtained soil from further afield so as to increase the variety of soil types seen. Children were encouraged to examine the samples, describing and comparing them.

Magnifiers were available beside the display for use if the children so desired. Children were asked to provide a picture and description of a few of the soil samples.

b. Testing soil for plant growth.

The emphasis of this activity was to find out something about how children might test soils for their suitability for plant growth. Ideas about how this might be approached were considered as important as the child's opinion about which soil would actually support plant growth the better.

Two different looking samples were shown to the children who were asked how they would find out which one was better for plants to grow in. Older children produced written accounts of their proposed investigations while younger children talked about what they would do.

2. What is under the ground?

This activity addressed the issue of what children consider to be underneath their feet and whether they are consistent in their views. In other words, do children recognise that once below the surface, the materials in adjacent places are the same. Few children have had experience of holes which go deeper than a few feet so this activity is asking children to draw upon their imagination and interpretation of information they may have come across in books, comics, television programmes of all kinds, as well as what people have said to them in the past. For many children a major problem might be separating fact from fiction.

Children were taken into the school playground and asked to imagine digging a hole down through the tarmac. They were encouraged to think about going down as far as they possibly could. They were then asked to record their thoughts as a drawing of what they would dig through and find in their hole. Children were also encouraged to say or show how deep different things were and how big or thick the materials were. Having completed this children were taken to a nearby area of grass and asked the same question and to produce another drawing; this time to show what they thought was underneath their feet when they stood on the grass.

3. Rocks

As with the activities centred on soil, this exploration experience involved the setting up of a display of rock samples to which the children had contributed. However, because the word 'rock' itself is interpreted in many different ways, it was suggested that the teachers first ask the children where they thought rocks might be found and what rocks look like. The responses to these questions could be written, drawn or be the summary of small group discussions. Children would then be encouraged to set up the display and to record their observations and reactions in an appropriate form.

4. Weathering in the environment

In order to explore children's ideas about changes in the environment, particularly those caused by the weather - the process of weathering - teachers were asked to show children examples of different items which had been affected by the environmental conditions e.g. damaged plants, including trees, church masonry and statues, tombstones, brickwork and wooden structures such as fences and outbuildings. The children were encouraged to examine the items, touching them where possible, and to draw pictures or write an account indicating how the item had got in that condition. A series of drawings could also be used to show what the item was like originally, what it was like at present and what it might be like in the future.

2.3.2 Pre-intervention Interviews - Further Elicitation

Members of the project team, including the advisory teacher, visited the schools to talk to children about their experiences during the exploration phase of the research. The interviews, which took place either in the classroom or in an otherwise unoccupied room, were conducted in an informal manner and every attempt was made to put children as much at ease as possible. In fact, in most classes, children were not only used to expressing their ideas but also extremely keen to do so.

Interviews in SPACE Project research have always been designed to extend the researchers' understanding of children's ideas emerging from the exploration experiences. That is, the interviews have been an attempt to probe and clarify the thinking that has emerged from the exploration experiences.

The interview schedule was not, however, drawn up until some feedback from the classroom activities had been received. This was obtained firstly through classroom visits when the progress of the activities could be seen and teachers could report what they had learned about the ideas of children in their classes. Secondly, some of the products from the activities, drawings, writing, etc., had been obtained and a preliminary scan was done to outline the thoughts they seemed to reveal.

The resultant interview schedule, which appears as Appendix IV, reflects the attempt to respond to the feedback received. Exploration experiences had elicited certain ideas which required further clarification. There was, however, no possibility of piloting interview questions.

The interview aimed to:

a. ascertain children's views on the nature and origins of soil;

b. identify the range of materials accepted by children as soil;

c. clarify the ideas expressed by children in their drawings of what was under their feet;

d. find out what children understand by the term 'rock' and to explore their ideas about the origins and permanence of rock.

In order to address these aims the interview was divided into four sections, each of which concentrated on one of the above aims.

In the first part of the interview, each child was shown a sample of soil which they were allowed to examine before being asked, 'What do you think soil is?' and 'What do you think is in it?'. If the child responded by referring directly to the sample then they were also asked what they might find in other soils. Questions about the formation of soil and its degree of permanence were then introduced. This line of questioning was pursued as far as was possible depending on the responses given by the child.

The second part of the interview was used to question each child about what they would accept as soil. This was done more specifically than previously. Five samples, A to E, were shown to the child who was asked to identify what was soil and what was not. They were asked to give reasons for their decisions. Interviewers were able to extend to questioning, if appropriate, to ask, for example, how sample A could be made into soil.

Children's own drawings were used to start discussions in the third part of the interview. Each child was asked to describe what they had drawn and the interviewer asked for clarification about such things as the scale and distances involved; the actual contents of a particular layer of material; whether there was anything underneath the bottom layer of the drawing. Any technical terms, such as 'core' or 'mantle', used by the children were probed in order to amplify what was actually meant. If the drawings indicated different structures under the playground from those shown under the grass, then the children were asked to give their reasons for the differences.

The final part of the interview concentrated on the children's ideas about rock. Two rock samples were used; one a jagged piece of limestone and the other a smooth piece of sandstone of approximately the same size. Initially, the child was asked to say what he/she thought when the word 'rock' was used and then the two samples were shown. For each sample, the child was asked whether they considered it to be a rock and why. Views about where rocks might be found were then elicited along with suggestions about how rocks are formed and the extent of them. As with all the interview sections, interviewers were able to extend the discussion if it was felt appropriate in order to clarify the child's ideas even further.

2.4 Weather research

2.4.1 Exploration Experiences - Exposure and Elicitation.

Weather is happening outside all the time so the initial phase of the research in which children are exposed to the concept area involved drawing their attention to what the weather was like from day to day. They were encouraged to make measurements and keep records of some aspects of the weather. Children were encouraged to relate the type of weather, its changes and ways in which it affected them. During the exposure phase, the teachers kept careful records about what the children said and did. These were then discussed at the pre-elicitation meeting and four main elicitation activities were designed and agreed upon. The instructions for these are given in Appendix V.

Children were asked to record the weather over a period of time but using symbols rather than words or pictures. The teachers talked to the children about their records and their reasons for the symbols chosen. The annotations were added to the children's records where further interpretation was needed.

As a result of the exposure experiences and the first activity, children had been alerted to the fact that the weather changed, sometimes dramatically in a very short space of time. They were then asked to produce a 'picture strip' set of drawings indicating why particular changes occurred.

The third activity involved children watching a television weather forecast and then being asked to say:

* what the forecaster was doing
* how he/she knows what the weather is like
* how he/she knows what the weather is going to be like tomorrow
* what the various symbols mean.

Children were asked to give their responses either in written form or orally, depending on their age and ability to put their ideas on paper.

In order to explore children's ideas about the effects of the weather on the environment, the children were asked to produce a piece of creative writing and drawings in which they had control of the weather. They were asked to describe the effects of the different types of weather they created and the changes they could cause in both the short and long term.

3. CHILDREN'S IDEAS

3.0 Introduction

This chapter outlines the ideas which were encountered amongst children prior to any intervention input. All data are drawn from the participating schools using a range of elicitation techniques. Where frequencies of the incidence of certain ideas are given, these refer to the stratified random sample of children who comprised a sub-set of the study group as a whole. This group of children participated in the in-depth interviews both before and after intervention. Table 3.1 shows the characteristics of the interview sample by age group, gender and achievement band.

Table 3.1 Sample by Age Group, Gender and Achievement Band (n=58)

		Infants	Lower Juniors	Upper Juniors	
Low Achievers	Girls	3	2	4	19
	Boys	4	3	3	
Medium Achievers	Girls	3	1	4	19
	Boys	4	4	3	
High Achievers	Girls	5	3	2	20
	Boys	1	4	5	
	27 Girls 31 Boys	20	17	21	58

Achievement band was defined by each class teacher's impression of every child's overall scholastic performance relative to the performance of the year group as a whole. This is a fairly loose definition but was found to be easily operable by teachers. It was emphasised that actual rather than potential performance should be used in allocating children to one of the bands, 'high', 'medium' or 'low'. Children were then randomly selected from these bands by the interviewers in order to avoid the sample being skewed by ability.

The identical sample of children was interviewed between six and ten weeks later, following the kinds of intervention activities described in Chapter Four. (Any child for whom either pre- or post-intervention data were missing as the result of absence or illness has been omitted from the analysis.)

The findings reported in this chapter describe the ideas which children revealed during the initial interviews. This phase is described as being 'pre-intervention', meaning before teachers have made any deliberate attempts to challenge, influence or inform children's thinking. (The term 'pre-teaching' is not used, since it would carry a different set of implications; the word 'intervention' more accurately carries the message of a whole range of impacts on thinking being possible). Having made this point, it is acknowledged that even a relatively passive activity such as asking a child's point of view is likely to provoke reflection and re-structuring of ideas, as that viewpoint is considered and articulated. Indeed, this acknowledgement is in recognition of the intrinsic drive of autonomous thinking which is harnessed by the constructivist approach to teaching and learning. The position of the research group was that the ideas being solicited during this phase of the project were **considered** points of view. Children were not ambushed and forced to comment on issues about which they had no view, or inveigled into saying what they did not mean. Chapter Five reports children's ideas after teachers' attempts to provoke reconsideration by a variety of methods, as compared to their ideas prior to the kinds of interventions described in Chapter Four. The particular interest is in any shifts in thinking which might be identified. Since the nature of the intervention was invented, discovered or shared as the result of (to some extent unpredictable) interactions which took place in classrooms, it should not be expected that the two sets of interviews, pre- and post-intervention, were identical. If a new insight emerged, it was pursued. Similarly, children's responses were always probed until the interviewer was satisfied at having understood the point of view which was being expressed, even if this took the interview into unexpected realms.

The data presented in this section avoid repetition of information which will be provided in Chapter Five. The specific issues which will be introduced are the following:

3.1 Ideas about the Nature and Origins of Soil

3.2 Ideas about Rocks

3.3 Ideas about Underground and the Structure of the Earth

3.4 Ideas about weathering

3.5 Ideas about weather

3.1 Ideas about the Nature and Origins of Soil

3.1.1 First thoughts

The researchers who conducted the interviews also visited classrooms from time to time and good working relationships were established with teachers. Consequently, the interviewer would be at least recognised by the children who were interviewed and the context of the discussion would have been familiar as the result of the teacher's introduction of the topic to the class. The result was that children tended to feel at ease during the interview and indeed were often eager and proud to discuss their thoughts. This enabled interviewers to turn to the issues germane to the project fairly comfortably and rapidly.

Following the practice of focusing interview discussions on actual materials or events whenever possible, the conversation about soil was opened by the interviewer presenting a small sample of soil on a sheet of white paper. The instructions to the interviewer were to commence by asking:

What do you think soil is?

followed by

What do you think is in it?

While at pains to establish the child's ideas in response to these specific areas of enquiry, interviewers were free to add non-directive encouragement and probes for clarification as they saw fit, according to the direction the child's responses took.

Table 3.2 Nature of First Thoughts Expressed about Soil*

	Pre-Intervention		
	Infants n=20	**Lower Juniors** n=17	**Upper Juniors** n=21
First response based on alternative name	50 (10)	23 (4)	5 (1)
First response based on use	40 (8)	29 (5)	19 (4)
First response based on origins	_	18 (3)	19 (4)
First response based on properties	10 (2)	12 (2)	14 (3)
First response based on mixed composition	_	6 (1)	24 (5)
First response based on location	_	_	5 (1)
First response based on another description	_	6 (1)	_
No response/don't know	_	6 (1)	14 (3)

*Percentages by age group. Raw numbers in brackets.

Table 3.2 records the nature of children's first response to the soil sample. Half the youngest group offered an **alternative name** as their identification of the material in front of them, a type of response which decreased steadily with age. Only ten per cent of the youngest age group offered an initial comment based on the properties of the soil sample. This more analytical type of response was much more common amongst the older children - 36% of lower juniors made an immediate comment about **origins, properties** or **composition,** this percentage rising to 57% for the upper juniors

The other major response category suggested by the infants related to the **function** of soil, with 40% of children describing the uses of soil. Once again, this kind of response declined with the increasing age of the sample.

Children's ideas about the nature of soil in terms of functions are summarised in Table 3.3 (Numbers are slightly enhanced as compared with Table 3.2 by those children who added to their initial definition during the course of the discussion).

Table 3.3 Thoughts about the Nature of Soil: Functions

	Pre-Intervention		
	Infants n=20	**Lower Juniors** n=17	**Upper Juniors** n=21
For growing plants	40 (8)	35 (6)	29 (6)
As a home for animals	5 (1)	–	5 (1)
Both of these uses	5 (1)	–	–
Other use	5 (1)	–	–
Don't know/no response of this kind	45 (9)	65 (11)	66 (14)

The most frequently suggested attribute of soil in terms of its use or function was 'for growing plants'. (There is a link here with the commonly reported idea that **all** plants' growth needs are derived from the soil in which they are rooted). The frequency of this kind of response was broadly in the 30-40% band, decreasing slightly with age. The other response which was encountered was offered by two infants and one upper junior (about 5% of the whole sample). This was the suggestion that soil has the function of being a home for animals.

> *It goes under grass. Worms live in it, and snails.*
>
> Y2 B L[1]

[1] A code is used to identify the characteristics of members of the individually interviewed sample as follows:

R	Reception	B	Boys	H	High Achiever
Y1	5-6 years	G	Girl	M	Medium Achiever
Y2	6-7 years			L	Low Achiever
Y3	7-8 years				
Y4	8-9 years				
Y5	9-10 years				
Y6	10-11 years				

You plant your plant in it. It helps it grow cos it sucks all the soil up and it helps it stay in one place.

Y3 B L

As reported in the comments relating to Table 3.2, younger children especially tended to identify soil by re-naming it, while older children tended to be more analytical. It is suggested that the quality of this difference in the nature of children's responses is progressive. It seems that the younger children are tending to re-name soil as a homogeneous material; their names are alternatives - 'mud', 'sand', 'stones', 'compost' or 'dust'.

Soil is a kind of dust.

Y2 G H

Table 3.4 Thoughts about the Nature of Soil: Alternative Names and Ideas about Composition

	Pre-Intervention		
	Infants n=20	**Lower Juniors** n=17	**Upper Juniors** n=21
Refers to soil as mud	25 (5)	12 (2)	5 (1)
Refers to soil as sand	15 (3)	–	–
Refers to soil as stones	–	12 (2)	–
Refers to soil as compost	5 (1)	–	–
Refers to soil as dust	5 (1)	–	–
References to other composition	5 (1)	18 (3)	5 (1)
Refers to soil as a mixture	15 (3)	6 (1)	38 (8)
No reference to soil composition	30 (6)	52 (9)	52 (11)

Some of the younger children perceived soil as a mixture, but this was more frequently the case with the upper juniors.

The geologist's definition of soil is as a mixture of inorganic and organic (including living) materials. An important concept is that soil is a dynamic aggregation of materials with a changing history and future. Both of these areas were explored with children. Three questions were included in the interview protocol to elicit children's comments on these issues. The nature of the questions was such that the manner in which the response was framed by children was unpredictable. For example, when asked:

Has the soil always been there on the piece of ground where it was found?

they might refer to origins, location or transformation. The way this was dealt with was to have available a cluster of three questions which addressed overlapping issues. (It was always the practice to code children's responses according to the issue under consideration rather than tightly tied to the specific question posed). The other two questions were:

How do you think soil got to be like this? and
Where do you think soil comes from in the first place?

Seventy-nine percent of the sample suggested some manner of change in the soil. The specific nature of those changes are summarised in Tables 3.5 and 3.6.

Table 3.5 describes the more superficial changes, especially those which could be attributed to many other materials.

Table 3.5 Ideas about Change in Properties of Soil: Appearance and Texture

	Pre-Intervention		
	Infants n=20	**Lower Juniors** n=17	**Upper Juniors** n=21
Soil Changes:			
Between wet and dry only	10 (2)	12 (2)	24 (5)
Between wet and dry\hard and soft	15 (3)	6 (1)	10 (2)
Colour	10 (2)	6 (1)	–
Colour and wetness	15 (3)	–	5 (1)
Between hard and soft only	5 (1)	12 (2)	–
Colour, wetness and hardness	–	–	5 (1)
No mention of above properties	45 (9)	64 (11)	56 (12)

Thirty-five percent of children mentioned the change in water content or wetness of soil, some of these responses being elaborated to include secondary effects such as texture. Water is an extremely important agent in soil change, in erosion and transportation for instance, so it is encouraging to have evidence that many children have generated a useful observational context. Table 3.6 includes other changes mentioned by children, some of which may be inferred to be closer to specific processes in soil formation and transportation as a geologist might describe them.

Table 3.6 Other Changes to Soil

	Pre-Intervention		
	Infants n=20	**Lower Juniors** n=17	**Upper Juniors** n=21
Growth in soil	20 (4)	–	19 (4)
Internal movement of soil by people	5 (1)	18 (3)	10 (2)
Internal movement of soil by small creatures	5 (1)	–	–
Internal adjustments of layers	–	–	10 (2)
Clumping of soil	5 (1)	6 (1)	10 (2)
Rock formation	–	6 (1)	5 (1)
Clumping - rock formation	5 (1)	–	–
Deposition of more soil	5 (1)	12 (2)	–
Enrichment by worms	–	–	5 (1)
Removal by wind	–	6 (1)	10 (2)
Removal by water	–	6 (1)	–
Other changes	10 (2)	18 (3)	14 (3)

The frequencies throughout are small, but it is of interest that they occur at all. For instance, it would be extremely useful for a teacher embarking on a class discussion in this area to know that such a range of relevant ideas is likely to be fed into the discussion by children themselves. The range of responses in Table 3.6 encompass some key geological processes, though considerations of plant growth and human activity predominate.

Once it is established that children are entertaining the idea of soil having dynamic properties, in contrast to a view that soil has always existed since the beginning of time,

I think since the world was made it's been there....

Y5 G M

the question of the origins of soil may be addressed. Children's ideas about origins are presented in Table 3.7.

Table 3.7 Ideas about the Origins of Soil

	Pre-Intervention		
	Infants n=20	**Lower Juniors** n=17	**Upper Juniors** n=21
Inorganic origin	5 (1)	12 (2)	19 (4)
Organic origin	–	6 (1)	–
Organic and inorganic	–	–	5 (1)
Produced by worms	–	–	5 (1)
Other origin	5 (1)	–	–
No reference to origins	90 (18)	76 (13)	57 (12)
No response\don't know	–	6 (1)	14 (3)

Only 19% of children (10% of infants, 18% lower juniors and 29% of upper juniors) were able to express some kind of theory about the origins of soil during the pre-intervention interviews. One child described both organic and inorganic antecedents and constituents:

R *It came from millions of years ago when the vegetation rotted away and it decayed over millions of years and mixed with sand and gravel and made soil.*

Q Where did the sand and gravel come from?

R *The sea and rocks. The sands at the bottom of the sea and seas washing it on to the shore and the winds blowing it inland. The wind is blowing the rocks and that's taking little bits of it which are broken down into little particles.*

Q What would you call those particles?

R *Sand or gravel.*

Y6 B H

That young children are confused by geological time is not surprising; this child's assumption is that all changes in soil would need to be described as having occurred over a period of millions of years. Such an assumption was not uncommon. It completely overlooks the fact that the processes described are still going on as a continuous mechanism: 'new' soil is a logical and realistic possibility.

R *It's like been a plant, over millions of years and then it's just got under ground and turned into soil.*

Q Well, how did that happen?

R *Well, over millions of years, it just starts to rain, under the ground, and it's like insects, beetles and things start to ... start...it's like re-cycling it really, into soil; and it turns right round. Worms get to it and then it goes, the birds eat the worms and then the bird dies and it goes rotted up into the soil. And it just goes round.*

Q I see. That's a very interesting answer, John. What do you think is in soil like that?

R *Things like vegetables and trees, roots, wings, feathers...all sorts, really.*

Y3 B M

3.2 *Ideas about Rocks*

3.2.1 *First thoughts*

The next section of the pre-intervention interview turned to a consideration of each child's understanding of the term 'rock' together with their ideas about the origins and permanence of rock.

The interviewer introduced the topic by asking:

What do you think of when I say the word 'rock'?

Children's first thoughts in response to this question are summarised in Table 3.8

Table 3.8 First Thoughts about Rocks

	Pre-Intervention		
	Infants n=20	**Lower Juniors** n=17	**Upper Juniors** n=21
Refers to physical properties	50 (10)	47 (8)	38 (8)
Mentions related words	5 (1)	18 (3)	33 (7)
Refers to where found	5 (1)	6 (1)	14 (3)
Refers to origins (how it is formed)	–	12 (2)	10 (2)
Refers to uses	5 (1)	–	–
Refers to what it becomes or changes to	–	6 (1)	–
Refers to composition	–	6 (1)	–
Other descriptions	–	6 (1)	5 (1)
No response/don't know	35 (7)	–	–

The dominant response referred to the **physical properties** of rocks and accounted for half of the infants and a gradually decreasing proportion of middle and upper juniors' responses. The kinds of properties to which children referred are illustrated by the following comment:

> *Rock is hard and it's grey and they have different colours of rock, white and black. Sometimes, when you throw them, some of them break but big hard ones don't.*

Y2 B M

Two responses which occurred at lower frequency levels but increased with age were references to related **vocabulary** and indications of **location**. For example, many children used the word 'stone', but for some, the distinction between what was rock and what was stone seemed to be important.

> *A rock is bigger than a stone. A stone is dead small, like that big* (indicated size by curling index finger.) *A rock is about as big as me.*

Y2 B L

When asked to indicate the size of a big stone, this boy made a circular form with both hands, finger tips touching.

The locations which children associated with rocks tended to be cliffs, mountains and the sea-shore, i.e place where rocks are exposed:

> *Mountains are made of rocks...sometimes people put the rocks on the mountains.*

Y2 G L

The reference to people putting the rocks on mountains could be an interpretation of the building of cairns used as monuments or landmarks which are so common in the British landscape, or even dry stone walls.

About ten per cent of children in the two older groups referred to **origins** of rock, in the sense of how it was formed.

Soil which has gone really hard. When its really hard, we call it rock. It goes hard when the sun goes down on it, like at the beginning of the Earth. We call it rock. Rocks are slowly being formed. The sun goes down on them and makes them go hard.

Y6 G H

3.2.2 Consideration of Rock Samples

The next stage in the interview involved the presentation of two samples of rock. Since their perceptible qualities were important to children in the sense that it was to these that they referred, rather than more abstract properties, the concrete nature of the samples will be described as well as a brief geological introduction. The first was a piece of limestone, greyish in colour, jagged and about 200 ml in volume. The second sample of rock was sandstone, roughly the same dimensions as the limestone, but brown in colour and smooth and rounded in form. In the event, the physical attribute of each sample appeared to be a determining or influential factor in children's decisions as to whether or not to treat a specimen as rock or not. Responses are summarised in Table 3.9.

Table 3.9 Responses to Viewing Two Rock Samples

	Pre-Intervention		
	Infants n=20	**Lower Juniors** n=17	**Upper Juniors** n=21
Jagged and rounded samples are rocks	45 (9)	29 (5)	43 (9)
Jagged sample only is rock	30 (5)	29 (5)	43 (9)
Rounded sample only is rock	–	6 (1)	5 (1)
Uncertainty shown	30 (6)	35 (6)	9 (2)

Less than half the sample at all ages accepted both specimens as being rock. Roughly one third of the interview sample (slightly more amongst the upper juniors) acknowledged only the limestone as being rock. Much rarer was the response which treated only the rounded form as rock (two children, both juniors). About one third of the children in each of the two younger age groups were too uncertain to offer a response.

In making their judgements about whether or not a given specimen was rock, many children used attributes which could be treated in a bi-polar fashion. Specifically, they made comparisons such as hard/soft (hardness), light/heavy (density), large/small (volume) and rough/smooth (texture). Unfortunately, trying to make sense of how children were using these attributes to classify the samples was complicated by the fact that they referred sometimes to the attribute as being present, at other times to the fact of its absence, in an inconsistent manner. Table 3.10 illustrates this point.

Table 3.10 Categorisation of Rock Samples by Rough/Smooth Attribute

	Pre-Intervention		
	Infants n=20	**Lower Juniors** n=17	**Upper Juniors** n=21
Rough so rock in one case, smooth so not rock in other	–	24 (4)	43 (9)
Inconsistent use of rough/smooth in categorising samples	10 (2)	12 (2)	24 (5)
Rough so rock in one case, no mention in other	15 (3)	29 (5)	5 (1)
Smooth so not rock in one case, no mention in other	15 (3)	18 (3)	5 (1)
Rough so rock in both cases	5 (1)	–	10 (2)
Other responses	5 (1)	6 (1)	5 (1)
No mention of form, texture or shape	50 (10)	12 (2)	10 (2)

The single most frequent category of response (22% of the sample) was the idea that the rough specimen (limestone) was rock but that the smooth specimen (sandstone) was not. None of the infants offered this defining distinction. Some children (12% of the sample) indicated that in their judgement, the sandstone was not rock because it was smooth, without suggesting that the limestone was rock because it was rough. Sixteen per cent of the sample used this attribute inconsistently, a lapse which was not restricted to the younger representatives of the sample. Indeed, inconsistent use of this criterion was most frequent amongst the upper juniors (24% of that age group).

Table 3.11 summarises the frequencies with which four commonly encountered bi-polar constructs were used by children in determining whether or not each of the two samples were rock.

Table 3.11 Use of Certain Perceptible Attributes in Classifying Specimens as Rock or Not Rock

	Infants n=20	Lower Junior n=17	Upper Juniors n=21
Rough/Smooth	50 (10)	88 (15)	91 (19)
Inconsistent use of rough/smooth	10 (2)	12 (2)	24 (5)
Hard/Soft	40 (8)	47 (8)	29 (6)
Inconsistent use of Hard/Soft	-	6 (1)	5 (1)
Large/Small	25 (5)	29 (5)	33 (7)
Inconsistent use of Large/Small	-	-	5 (1)
Light/Heavy	30 (6)	24 (4)	29 (6)
Inconsistent use of Light/Heavy	-	-	-

Rough/smooth was offered the most frequently and with the most inconsistency. Light/heavy (whether alluding to mass or density is uncertain) was used least frequently but most consistently. It is difficult to make any firm sense of these data; it might be safest to conclude that the children themselves appear to be struggling. Firstly, the criteria themselves refer to features which are relatively superficial. They are not, for example, science-specific, far less geology-specific in nature. Secondly, in using the attributes in a categorical rather than scalar fashion, comparisons are inevitably crude. Perhaps the use of two samples encourages such a limited response. Instruments which might have promoted closer inspection or some form of investigation (hand lenses, files, etc) were not provided.

The data suggest that children might be encouraged to make closer and more detailed observations - a conclusion which repeats the experience of much of the SPACE project research with this age-group. They might be encouraged to take measurements and attempt to quantify their criteria, using a range of specimens rather than just one or two. (Such recommendations are made with the benefit of a hindsight based on a later availability of data summaries than was the case when the intervention guidelines were framed with teachers).

Table 3.12 reports the most frequently encountered criteria children used to categorise the samples as rock or not rock, other than bipolar attributes.

Table 3.12 Use of Non-bipolar Criteria in Categorising Samples as Rock or Not Rock

	Pre-Intervention		
	Infants n=20	**Lower Juniors** n=17	**Upper Juniors** n=21
Colour indicates one sample rock, no mention in other	5 (1)	_	19 (4)
Looks like rock in one case, no mention in other	10 (2)	12 (2)	0
Doesn't look like rock in one case, no mention in other	15 (3)	_	5 (1)
Feels like rock in one case, no mention in other	5 (1)	12 (2)	_
Wrong colour for rock in one case, no mention in other	_	12 (2)	_
Colour indicates rock in both cases	_	6 (1)	_
Colour indicates one sample rock and other sample not rock	_	_	5 (1)
Looks like rock in both cases	5 (1)	_	_
Sounds like rock in one case, no mention in other	_	6 (1)	_

As with the criteria reported earlier, those in Table 3.12 tend to be of the general nature that might be used to describe any object; they refer to surface characteristics and include judgements of an intuitive nature. References to colour predominate, suggesting that there are right and wrong colours for rock. The source of such stereotypes is uncertain; it is clear that they were imported by the children rather than suggested by the line of questioning. It may be the case that these responses are evidence of children's lack of any considered linguistic or analytical frameworks for discussing criterial attributes of rock. Rocks and their attributes are not everyday topics of conversation; on the other hand, fairly well-defined vernacular definitions appear to be in operation which children use to distinguish between rock and stones or pebbles. This semantic distinction was behind the rejection of the rounded specimen as a rock on the part of a large proportion of children, as revealed by the data in Table 3.13.

Table 3.13 Assumptions about the Logical Relationship between 'Stone' or 'Pebble' and 'Rock'

	Pre-Intervention		
	Infants n=20	**Lower Juniors** n=17	**Upper Juniors** n=21
The rounded sample is a stone/ pebble and not a rock	15 (3)	29 (5)	48 (10)
The rounded sample is a stone/ pebble and is a rock	20 (4)	12 (2)	5 (1)
Both samples are stones/pebbles and rocks	5 (1)	6 (1)	5 (1)
Both samples are stones/pebbles and not rocks	5 (1)	6 (1)	5 (1)
The jagged sample is a stone/ pebble and not a rock	10 (2)	–	–
The jagged sample is a stone/ pebble and is a rock	5 (1)	–	–
Other response	–	6 (1)	–
No mention of stone/pebble	40 (8)	41 (7)	38 (8)
Total treating stones/pebbles as instance of rock	30 (6)	18 (3)	10 (2)
Total treating stone/pebble as non-instance of rock	30 (6)	35 (6)	52 (11)

Even in adult usage, there are semantic complexities to unravel in the usage of the words 'stone', 'pebble' and 'rock'. Conventionally speaking, both stones and pebbles would be regarded as fragments of rock; generally, when used to describe discrete objects, it would be highly likely that 'stones' and 'pebbles' would carry connotations of size. For example, it might be generally agreed that both stones and pebbles are capable of being thrown by a person. (Anything unmoveable might be more likely to be described as a 'boulder'). This definition breaks down with respect to particular stones such as monumental or neolithic 'standing stones'. Usually, a 'stone' is a more generic concept than a 'pebble': pebbles are stones of a particular shape, being smooth and rounded. As well as describing discrete objects, the word 'rock' is used to describe the material in general. When used in the singular form without a definite article, 'stone' also is used as a collective noun referring to a class of material. It is unlikely that these definitions would ever have been articulated in the experience of any of those children who were randomly selected for interview. That their responses implied a certain amount of confusion about the logical relationships between the various terms is, on the basis of this analysis, unsurprising.

The use of 'stone' and 'pebble' is not distinguished for the purposes of Table 3.13, the interest being whether, whichever term is used, it is assumed to be a sub-set of the class of objects made of rock. Overall, there are two related trends clearly in evidence. Firstly, the overall number of children treating stones and/or pebbles as instances of the class of material known as rock was found to **decrease** with age; the logical separation of the terms 'stone' and 'pebble' from 'rock' was found to **increase** with age. These trends are in the opposite direction to what would be regarded as educational progress.
Some attempt to unpick how children were distinguishing the words 'rock' and 'pebble' was also attempted (see Table 3.14).

Table 3.14 Use of 'Stone' and 'Pebble' to Describe Two Rock Samples

	Pre-Intervention		
	Infants n=20	**Lower Juniors** n=17	**Upper Juniors** n=21
'Stone' used for rounded sample only	45 (9)	24 (4)	38 (8)
'Pebble' used for rounded sample only	15 (3)	24 (4)	19 (4)
'Stone' used for both samples	10 (2)	18 (3)	10 (2)
'Stone' used for jagged sample only	15 (3)	6 (1)	–
No use of 'pebble'	85 (17)	76 (13)	81 (17)
No use of 'stone'	30 (6)	53 (9)	52 (11)

Table 3.14 tends to confirm the use of 'stone' as being used more inclusively than 'pebble': there were no instances of the jagged piece of limestone being described as a pebble, though the rounded sandstone was very frequently referred to as a 'stone'.

Q Have a look at this one. (Rounded sandstone specimen.) Do you think that's rock. Is it a piece of rock?

R *It's like a stone really, but you could call it rock.*

Q Why? How do you decide?

R *The weight. If it's hard or not. (Taps sample). What type of noise it makes. The shape.*

Q What shape does rock have to be?

R *No shape! It can be any shape, but this one's like a stone because of its shape. It's like, you know, stones have a curve all round it and the fossil* (referring to limestone specimen) *has like bits broken off it so it's been eroded.*

Y3 B M

This boy could not be said to be resolute in his definitions of 'stone' and 'rock'; the uncertainty he acknowledged was not an expression of personal doubt so much as an assertion of a fuzzy set boundary. It is also interesting to note that he is defining a stone as being rounded in form. Furthermore, the rough and irregular appearance of the limestone is taken to be evidence of erosion. The implication seems to be that the rounded form is **not** understood to be the result of erosion. It is not too difficult to empathise with a notion that rounded rock forms are in some way discrete and complete, natural phenomena; that just like conkers or mushrooms, they have a particular shape. The relative symmetry, precisely the result of erosion processes, is misinterpreted.

Some attempts to categorise the two rock samples had more geological relevance. For example, as indicated in Table 3.15, a small number of children - only four (7 % of the total sample) - ventured more specific names.

Table 3.15 Other Categorisations of Rock Specimens

	Pre-Intervention		
	Infants n=20	**Lower Juniors** n=17	**Upper Juniors** n=21
The rounded sample is rock and name of rock ventured	5 (1)	12 (2)	_
The jagged sample is a fossil and is rock	_	6 (1)	5 (1)
The jagged sample is a fossil and is not rock	_	6 (1)	_
The jagged sample is rock and name of rock is ventured	5 (1)	_	_

Three children identified the limestone as fossil material, one expressing the belief that as such, it was not rock.

Q Now, would you call this rock?

R *No. Fossil*

Q Tell me more.

R *Well, there must have been something like eggs here, because there's like bumps on it. This would have been on a cliff in the dinosaurs' age and what would have happened was something's happened to the dinosaurs that scientists are looking at.*

Y3 B M

For most children, the identification of the samples as rock involved a catalogue of criteria which were or were not met. Table 3.16 summarises the number of criteria mentioned by children as they reached their decisions.

Table 3.16 Number of Attributes Referred to in Samples

	Pre-Intervention					
	Inf n=20	**LJ** n=17	**UJ** n=21	**Inf** n=20	**LJ** n=17	**UJ** n=21
No. of attributes:		**limestone**			**sandstone**	
4	–	–	14 (3)	–	12 (2)	5 (1)
3	15 (3)	18 (3)	14 (3)	–	–	5 (1)
2	30 (6)	18 (3)	24 (5)	25 (5)	12 (2)	10 (2)
1	45 (9)	29 (5)	29 (6)	15 (3)	6 (1)	19 (4)
0	5 (1)	–	10 (2)	5 (1)	18 (3)	5 (1)
Not classed as rock or not applicable	5 (1)	35 (6)	10 (2)	55 (11)	53 (9)	57 (12)
Mean number	1.5	1.2	1.8	0.7	0.8	0.7

There were no clear age trends in the number of attributes children articulated, but more features were indicated in relation to the limestone than was the case with the sandstone. This may be because they found the judgement more problematic; it may have been because they were more certain; it could even have been an order effect. In contrast, responses suggesting that the samples were **not** rock tended to be more brief and categorical.

3.2.3 The Location and Extension of Rock Sources

The interview turned next to a consideration of where children thought rocks might be found. This was introduced by asking:

Where can we find rocks?

This question set the scene for the issue of particular interest, children's ideas about the existence and extent of bedrock. While in global geological terms, that proportion of rock which is visible above sea level and is not covered by alluvial deposits and vegetation is a tiny proportion of the whole, the suspicion was that for children, the perception was the reverse: rock would be treated as that material which tends to be visible in fragmentary form, or in larger masses in specific locations only. Thus, the supplementary question:

Can we find rock under where we are now?

had far-reaching implications for tapping children's basic understanding of a fundamental geological concept relating to the structure of the Earth. Responses are summarised in Table 3.17.

Table 3.17 Thoughts on the Possibility of the Presence of Rock Underground

	Infants n=20	**Lower Juniors** n=17	**Upper Juniors** n=21
Yes, there's rocks in the soil (under here)	50 (10)	47 (8)	29 (6)
Yes, there is rock under everywhere/a layer of rock	5 (1)	6 (1)	38 (8)
Yes, in building foundations	_	18 (3)	19 (4)
Yes, but no reason offered	10 (2)	18 (3)	_
Total positive responses	85 (17)	94 (16)	86 (18)
No, rock makes a surface bumpy	_	_	10 (2)
No, but no reason offered	10 (2)	_	_
No, no rocks, just stones	5 (1)	_	_
No response	_	6 (1)	5 (1)

Overall, 88% of the sample responded positively to the question, indicating a belief in the presence of rock under the place where the interview was taking place. However, it is probable that for the majority of children, their idea was not equivalent to the notion of bedrock. About half the children at the two younger ages and about one third of the upper juniors suggested the presence of rocks in the soil. This seems to imply that soil is the predominant material, within which is found a scattering of rocks. When some children indicated the presence of rocks, they were thinking of the foundations of the building. Rock and concrete were sometimes used as synonyms. The responses which might be consistent with a notion of bedrock were at the level of 6% amongst the lower juniors and 38% amongst the upper juniors, suggesting that the developmental trend is in the right direction, educationally speaking.

Ideas about how long rocks have existed were also probed and are summarised in Table 3.18.

Table 3.18 Thoughts on the Length of Time Rocks Have Existed

	Pre-Intervention		
	Infants n=20	**Lower Juniors** n=17	**Upper Juniors** n=21
Since the world began	–	24 (4)	24 (5)
A long time (forever)	20 (4)	–	10 (2)
Millions of years	5 (1)	18 (3)	10 (2)
Thousands of years	5 (1)	6 (1)	5 (1)
Hundreds of years	–	12 (2)	19 (4)
Tens of years	10 (2)	6 (1)	19 (4)
A few years	10 (2)	12 (2)	5 (1)
A year or less	5 (1)	–	–
A long time (not clearly forever)	40 (8)	18 (3)	14 (3)
Since a (specified event)	15 (3)	6 (1)	19 (4)
No response/don't know	5 (1)	–	5 (1)

The age of the Earth is usually estimated to be about 4.5 thousand million years, a number that most adults would find impossible to recall and almost as difficult to conceptualise. Children's ideas about the age of rock were expressed in a variety of ways. About a quarter of both middle and upper juniors used the idea of the beginning of the Earth, which they might have conceptualised in creationist or geological terms. The younger children especially (40% of infants) found it difficult to articulate any kind of scale or reference point, responding only in the generalised sense of 'a long time'; for an additional small group, further questioning clarified that the intended sense was 'forever'.

Other children attempted a quantified response in terms of millions, thousands or hundreds of years. About ten per cent of the sample seemed to have the idea that the presence of the rock was a relatively recent event - within a 'few years' or even a year or less.

Another group of respondents can be identified as using a specified event as a reference point. The time scales involved are not easily handled by young children and it is not entirely unexpected that there is evidence of mental struggle or confusion producing idiosyncratic replies:

Since people go camping

R B L

Dinosaurs and cave men were other reference points which children used as events to locate a particular historical period.

R *God might have made stones*

Q Have they been here for a long time, then?

R *I think so, yeh.*

Q How long?

R *(Laughs)*

Q Forever, or...

R *...uh, not forever. When God was made, when God was born, he made the world and when he made the world, he thought, 'There must be people to make things'. He might have made the rocks because...He needed help*

Q He made the rocks, did he, or did people make the rocks?

R *The people made the rocks and he made the rocks because they needed help, because, if they didn't have help it would take millions of years to do it. Because there was never...uh, no such things as machines there, then. And they must have done it with their hands, then.*

Y2 B H

It might be tempting to think of the creationist explanation of the origins of rocks as being simpler and more accessible to children than the geological version of events. The above transcript suggests otherwise. This boy is struggling with the notions of time and instantaneous creation. His attempts to resolve his problems lead him to an anthropocentric rationalisation.

3.2.5 Ideas about how Rocks change.

From the human perspective, geological changes tend to be imperceptibly slow in their effects. There is not much immediately obvious scope for experiential learning. Yet if children are to have any real understanding of the Earth's geology, they must somehow engage with the idea of change. Discussions about time, which is so often associated with changes in the nature of materials, led to a specific question about changes in rocks. Table 3.19 summarises children's ideas about how rocks might change.

Table 3.19 Thoughts About How Rocks Might Change

	Pre-Intervention		
	Infants n=20	**Lower Juniors** n=17	**Upper Juniors** n=21
To smaller pieces	20 (4)	59 (10)	71 (15)
Change location	10 (2)	12 (2)	14 (3)
Bits will stick together	–	18 (3)	10 (2)
Will turn soft/muddy	5 (1)	–	10 (2)
Will change colour due to rain	5 (1)	–	5 (1)
Will change colour due to sun	5 (1)	–	–
Will turn hard (with age)	–	–	5 (1)
Other changes	30 (6)	–	24 (5)
Don't know	–	6 (1)	–

The idea of rock breaking into smaller fragments increased steeply in frequency of occurrence with age to a level of 71% amongst the upper juniors. Some other physical processes which underpin an understanding of geological change were also in evidence, though with relatively low frequencies. A few children of all ages commented on changes of location. Aggregation or 'bits sticking together' was mentioned by 13% of the juniors. The effects of weathering by water and the sun were identified, though not the effects of freezing and glaciation or pressure and consequent generation of heat. In short, simple physical processes were identified with fairly modest frequencies. From a teacher's point of view, the position might be adopted that at least there are some starting points in evidence, and this is very encouraging.

The number of changes identified by the different age groups are summarised in Table 3.20, indicating the increasing ability to suggest changes with increasing age.

Table 3.20 Number of Rock Changes Mentioned

	Pre-Intervention		
	Infants n=20	**Lower Juniors** n=17	**Upper Juniors** n=21
0	30 (6)	29 (5)	14 (3)
1	60 (12)	53 (9)	43 (9)
2	10 (2)	18 (3)	33 (7)
3	–	–	10 (2)
Mean number	0.8	0.9	1.4

Table 3.21 records the specific agents of change nominated by children. The impression, alluded to earlier, of children's generally anthropocentric view about the place of rocks in the world is reinforced by the role the interview sample attributed to people in bringing about changes in rocks.

Table 3.21 Agents Which Bring about Rock Change

	Pre-Intervention		
	Infants n=20	**Lower Juniors** n=17	**Upper Juniors** n=21
People	35 (7)	24 (4)	24 (5)
Water/rain	10 (2)	–	24 (5)
Sea erosion	–	12 (2)	10 (2)
Rocks pressing down	–	6 (1)	14 (3)
Ageing	5 (1)	6 (1)	10 (2)
Wind	5 (1)	6 (1)	–
Sun/Heat	5 (1)	–	5 (1)
Ice/frost/glaciers	–	–	5 (1)
Other agents	10 (2)	6 (1)	10 (2)
Don't know/no response of this kind	15 (3)	41 (7)	–

A quarter of the upper juniors mentioned a role for water, including precipitation; in addition, a small number of juniors mentioned the erosion effects of the sea. Again, there is evidence of a very thin scattering of important geological ideas. The incidence of such suggestions across the whole sample is summarised in Table 3.22.

Table 3.22 Number of Agents of Rock Change Mentioned

	Pre-Intervention		
	Infants n=20	**Lower Juniors** n=17	**Upper Juniors** n=21
0	45 (9)	53 (9)	14 (3)
1	50 (10)	35 (6)	76 (16)
2	–	12 (2)	5 (1)
3	–	–	5 (1)
4	5 (1)	–	–
Mean number	0.7	0.6	1.0

3.2.4 Ideas about the Earth's structure

In addition to the questions posed during the individual interviews, all children in the classes which participated in the research programme were invited to make drawings of what was under the ground. They were asked to think of two different starting points for their drawings, the playground and a field. The quantified details of how the interview sample responded pre- and post-intervention are reported in tabular form in Chapter Five. In fact, children were very consistent in the important characteristics which they portrayed in their drawings, whatever the notional starting point. Some of the important qualities of their drawings are discussed in this section, any shifts in ideas following the intervention being reported in Chapter Five.

The characteristics of drawings which seem to have relevance to an analysis of children's understanding and which also appear to be age-related are as follows:

i) Whether drawings denote layers or relatively homogeneous material underground

ii)	The extent to which plant and animal activity and presence, as well as human activity and artefacts are a pre-occupation
iii)	When layers are represented, whether these are horizontal (local) or curved (global)
iv)	Whether layers of continuous rock are a feature
v)	Whether drawings represent a section through to the centre of the Earth and perhaps beyond
vi)	Awareness of distances and scales of subterranean features
vii)	Use of technical terms
viii)	Awareness of temperature differences within the Earth

No Layer Drawings

The 'no-layer' drawings were strongly associated with the younger children in the sample; **all** the lower and upper juniors in the interview sample included layers in their drawings. In the first drawing below, the child has put various objects in compartments, incidentally leaving out the geological features, other than "some mud at the bottom", together with a patch of stones. There are three references to animal material in the form of worms, insects and a bone, but no mention of plant material.

Figure 3.1

The next drawing does not show layers; the animal evidence includes live rats (at a depth of 40 feet) and bones. Pipes are shown as well as 'tar and stones'. The 'core' is drawn in the shape of an apple core (a confusion that was not uncommon) at an estimated depth of 90 feet. The unknown underground is, in short, populated with what is known and within what might be thought of as a domestic rather than global scale of distances. Whether the distances indicated in the drawing by this child - 2 feet, 10, 20, 30, 40 and 90 feet - actually mean anything is unclear. It is surprising that the measurement used is imperial rather than metric, perhaps inflluenced by usage at home rather than in school.

Figure 3.2

Drawings with Horizontal Layers

Figure 3.3 reproduces a drawing which delineates horizontal layers under the playground or 'yard'. The drawing is unusual in the number of technical references to minerals. It starts with 'tarmac' at the surface; the next layer is of soil which according to the child, occupies a depth of four or five metres. Next comes a layer of 'rocks and fossils' with intrusions of chalk and gold. There is a continuous layer of copper and of tin. These are followed by a layer of clay and another layer of rocks and fossils. More difficult to interpret is the layer labelled 'carbondioxide'. The penultimate layer is of lava which is adjacent to 'the core of the Earth', the latter drawn consistent with other layers in being shown as horizontal.

Figure 3.3

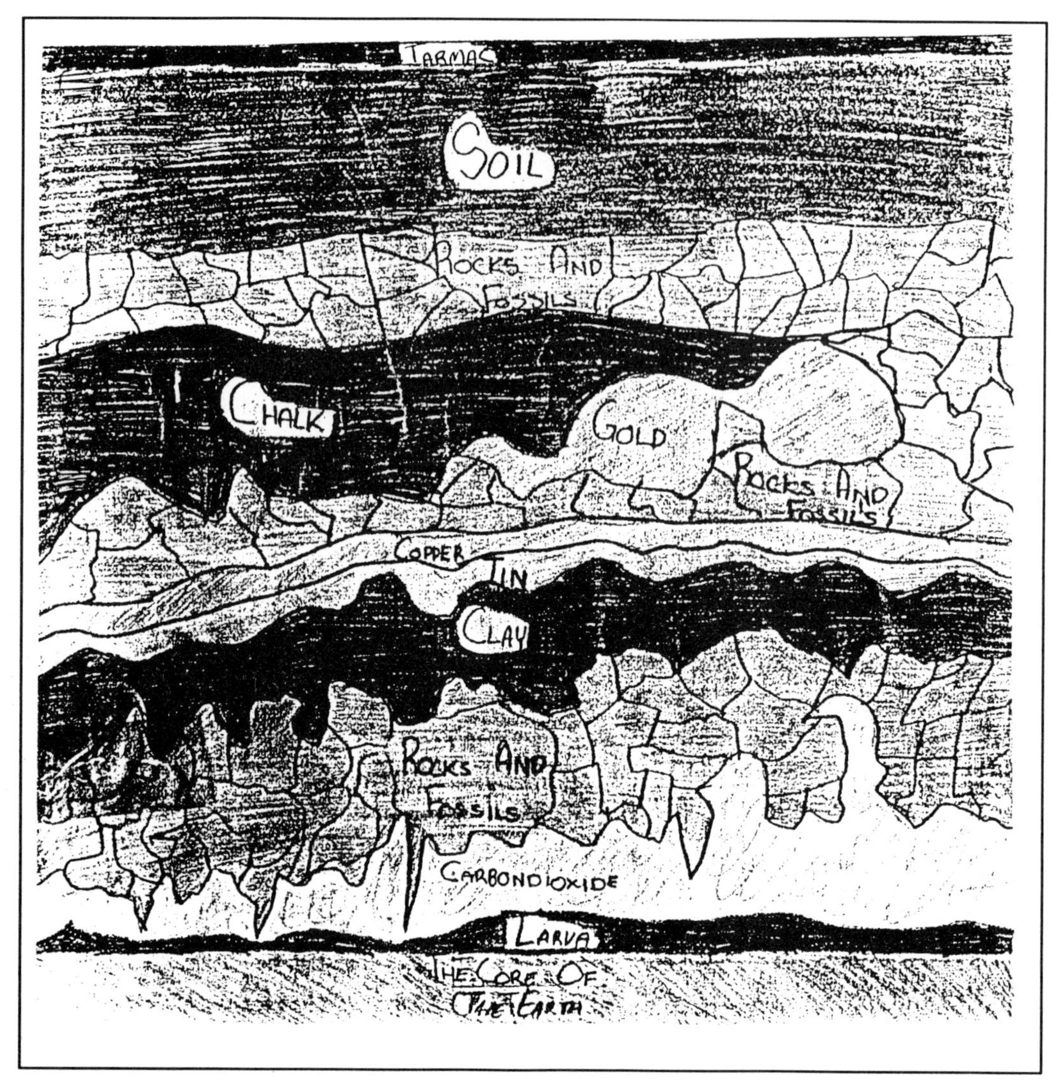

As in Figure 3.1 and 3.2, what is drawn is that which is familiar, but sited and reconstructed in an unfamiliar environment. For example, this child clearly has knowledge of a range of rocks and minerals occurring underground. There are also technical terms such as 'lava' and 'core' being used with some degree of accuracy, albeit not total. Perhaps there are indications here that, given the right support, such ideas are not inaccessible to children in the primary phase. There is a parallel to be drawn with charts of the Earth's surface features: the shapes of continents and oceans; contours of land masses and the depths of the seas; political borders. The conventions by which these features are represented have to be learned if atlases are to be interpretable. They are not encountered in any intuitive sense, but children are generally encouraged to begin by mapping the familiar environment, gradually extending to the unfamiliar. Perhaps this is the way for teaching/learning interventions to proceed, if it is agreed that knowledge about the internal structure of the planet we inhabit is worthy of early acquisition. That is, it might be appropriate to start with considerations of top soil and sub-soil, taking advantage of any local excavations or other features which expose what is immediately beneath the surface. These might include quarries, railway cuttings, sea cliffs, mines, and so on - all opportunities for children to see exposed what would otherwise be hidden from view underground.

Figure 3.4

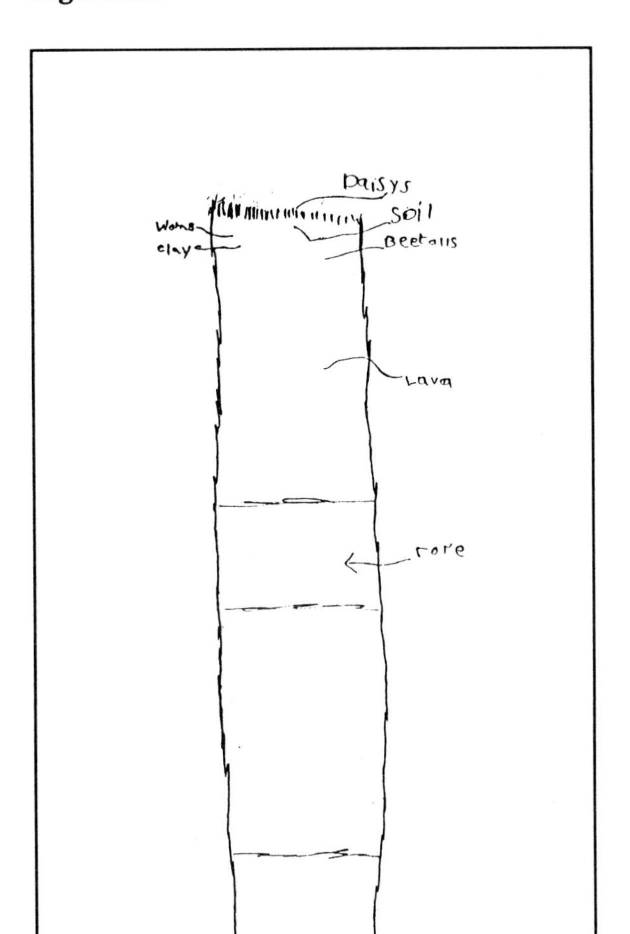

Figure 3.5

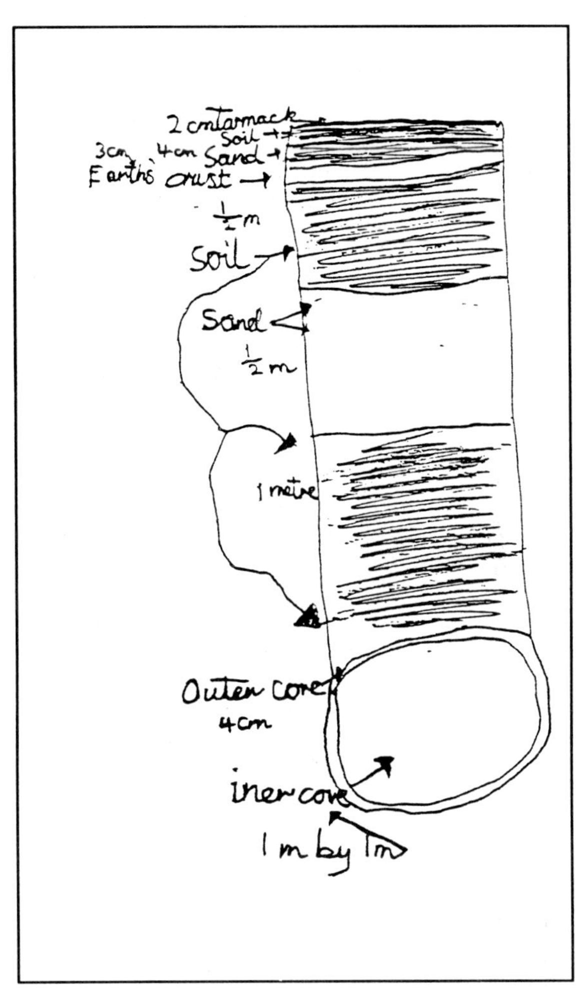

In Figure 3.4, the child has chosen to represent a complete symmetrical section through the planet. The drawing has some features characteristic of the younger children's responses: worms and beetles are labelled. There are also technical properties: 'core' and 'lava' are indicated. Although this drawing lacks much of the detail of some others, it shows a maturity in its grand conception.

Figure 3.5 offers an interesting comparison in that it reveals a transition between the representation of layers as horizontal and as circular; it is as if, by the time the drawing reached the core, the necessity of accomodating the spherical shape of the planet became conceptually irresistable. This drawing also serves as a reminder of how far some children were from comprehending the scale of what they were attempting to represent.

Figure 3.6

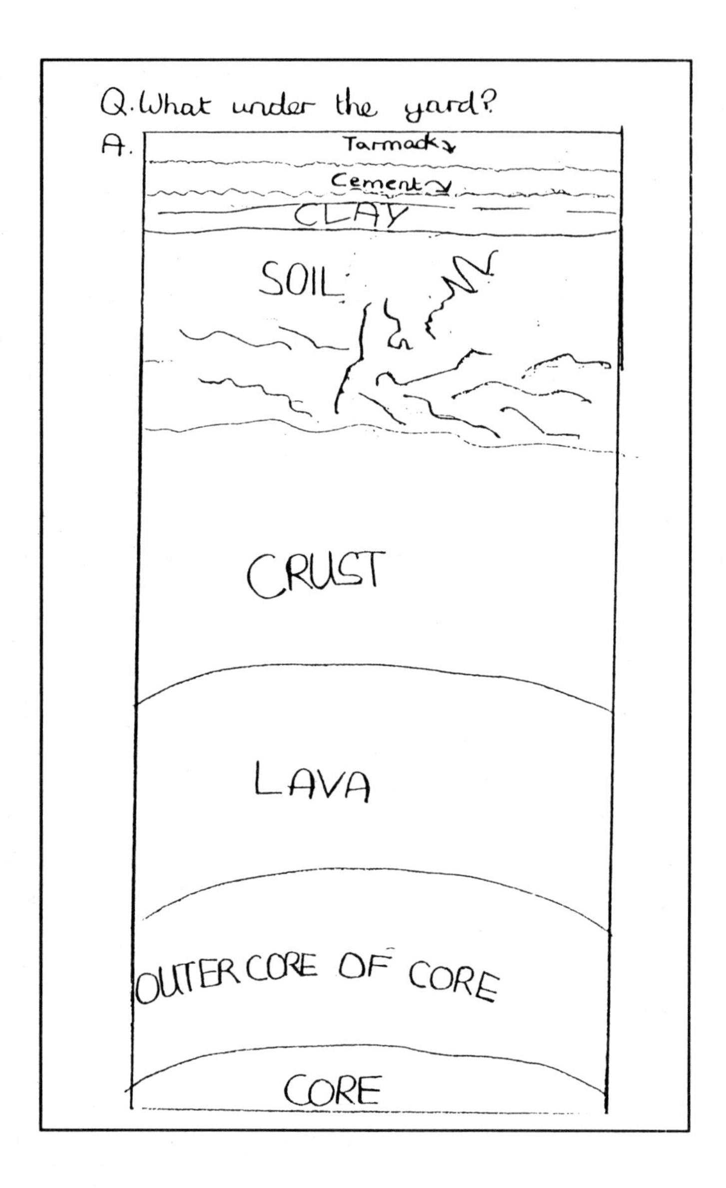

Figure 3.6 represents what appears to be a more assured reconciliation of the apparently flat surface of the Earth as we see it and the knowledge of the spherical shape of the planet. As in Figure 3.4, the grand design seems to be at the expense of any indication of detail. Perhaps this ability to conceptualise on a large scale **requires** that the fine detail be ignored, at least temporarily.

3.4 Ideas about weathering

The aim of this part of the research was to explore children's explanations of how weathering occurs in their local environment - i.e. within generally urban and suburban settings.

Children observed the effects of weathering on tombstones, brickwork, paintwork, and railings. Although many children considered that different weather conditions were the primary cause of the deterioration they had noticed, other explanations of weathering were offered by some. While some children cited a single cause for the weathering effects, others indicated that a combination of factors was responsible. Although children readily suggested causes of weathering, few attempted an explanation of weathering **processes** or **mechanisms.**

A commonly expressed interpretation of the weathering changes observed was that people must be responsible - an anthropocentric view. In some instances, it seems likely that ideas about the **agent** of change subordinated any possible **temporal** interpretation. This was the case when children explained the effects of weathering not as gradual but as a result of damage caused by people. This is another example of children having difficulty with the time scales involved in Earth Science phenomena. Of course, the situation is confusing to children because in some instances they are correct in assuming a human agency in relation to wearing paths and steps over long periods of time by the action of walking over them. These children, unlike the 'human damage' group, demonstrated some awareness that changes in the appearance of the stone was a gradual process which occurred over a long period of time. However, there is an associated danger in drawing children's attention to such outcomes that the attribution of change to a human agent may be over-generalised. The following example combines an awareness of the effects of human activity **as well as** natural effects.

Figure 3.7

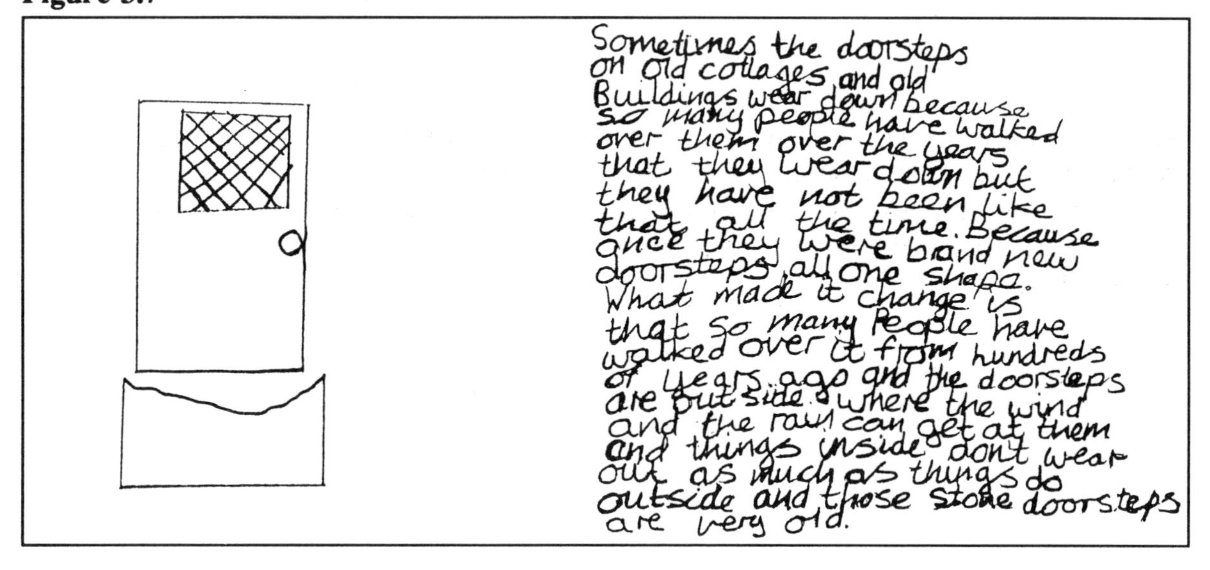

While many children referred to different weather conditions in **association** with weathering effects, explanations or hypotheses in terms of **processes** and **mechanisms** were less frequently encountered. Rain was described as causing weathering in a number of ways. Some children considered that the characteristics of new materials, such as being 'shiny', were washed away by rainfall. Others described rain as transforming the rock into a softer material. Weathering occurred as a result of this change in the property of rocks. The same child who described the wearing of the doorstep, Figure 3.7 above, had this to say about a gravestone:

Figure 3.8

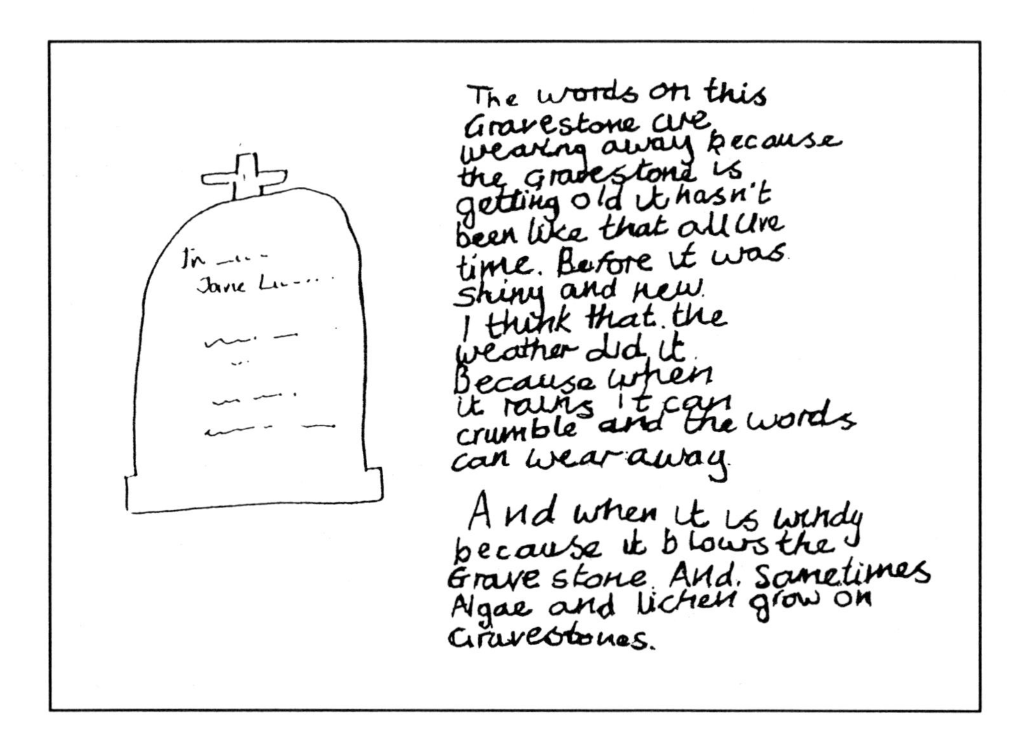

In everyday usage, the ageing process, 'getting old', is often treated as synonymous with 'deteriorating'. (The patina of antiques and the mellowing of fine wines are exceptions to this rule and not generally the concern of children). In a biological context, 'ageing' has very definite associations with becoming less functional. Children may be tempted to offer age alone as a causal effect, but as in the example above, with further reflection may manage to be more specific.

Children may be aware of the effects of the sun in contexts such as the fading of colours in classroom displays and the discolouring of newsprint. Some children perhaps over-generalised such effects in attributing the erosion of lettering on tombstones to the sun, suggesting that the sun had caused the lettering to fade.

Figure 3.9

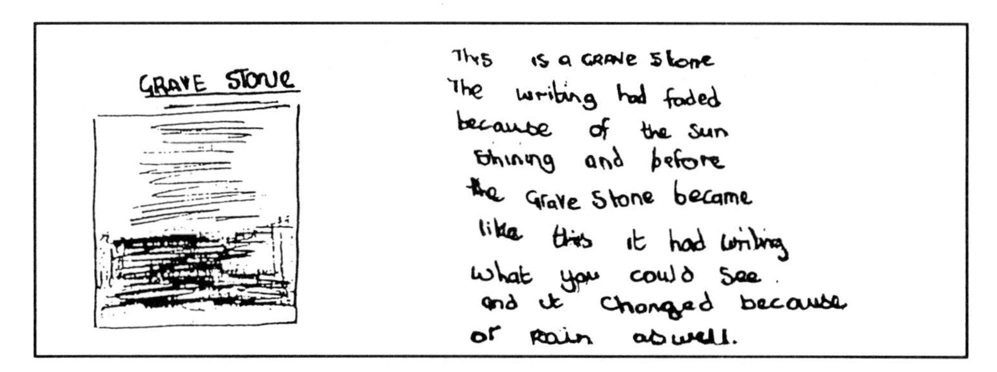

Flaking paintwork was a readily observable effect of weathering and one that children frequently encounter. Despite this example of weathering being within children's experience, many considered that the sun was solely responsible. Children tended to explain that the sun had heated the paint and caused it to flake.

Figure 3.10

A few children offered a combination of factors as causal agents in the weathering process. Some explanations were in terms of how the wind and rain and human activity had affected objects. The wind and rain were often cited together as causing weathering.

Figure 3.11

3.5 *Recording Weather Changes*

As indicated in Chapter One, the exploration of children's ideas about the weather was conducted at a different time, and with a different group of teachers, from the work on rocks and soil. The work on weather done in classrooms was not supplemented by individual interviews and was not quantified. Consequently, the descriptions of ideas are more tentative and more impressionistic in nature, pointing the way to further enquiry rather than attempting to draw firm conclusions.

3.5.1 *What Aspects of the Weather do Children Notice?*

It seemed that younger children tended to make only one record of the weather for a particular day. They may not be aware of how weather conditions vary during a day and may be unaware that recordings could be made at different points in the day in order to indicate the changes in the weather.

Young children's records of the weather typically contained such weather features as the sun, rain, clouds or wind. Some children record a windy day by a series of grey shadings, others may represent the wind in a human form. A few children may refer to the temperature in their records. They may describe changes in temperature in personal terms using phrases like, "it is boiling".

Figure 3.12

Many upper junior children demonstrated an ability to generate symbols which they use to describe particular weather conditions. Sometimes these symbols are accompanied by a written description of the weather. Children in this age group seemed to be increasingly aware of the possibility that weather conditions may change during the day and may decide to make more than one record of the weather on each day. Upper juniors seemed to be more aware of some of the ways in which a number of weather features may change. For instance, they might suggest that the clouds have changed colour to black and also that the sun may not be as bright. Symbols for each feature may well be depicted as overlapping, suggesting that there is an interrelationship between the change in cloud cover and the amount of sun.

Figure 3.13

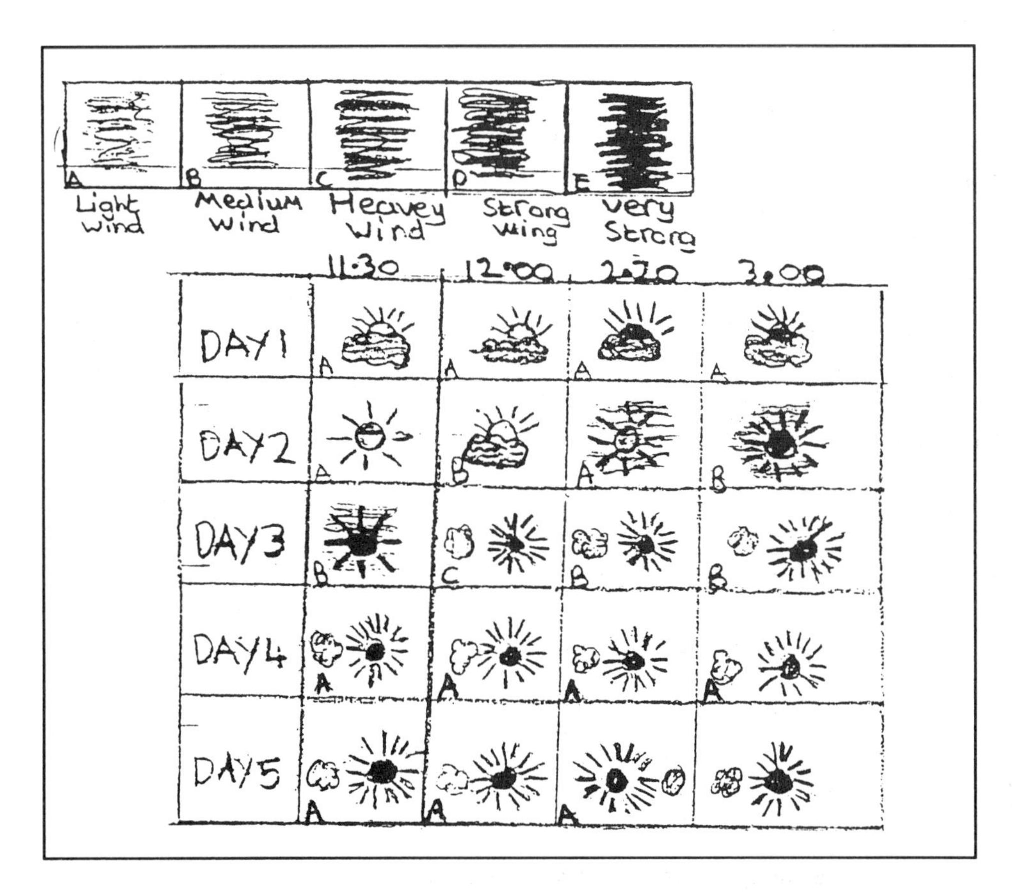

Junior school children showed an ability to record the finer variation in weather types in their symbolic representations of weather. They used a greater variety of ways to describe differences in the sun than did infants. Some used different shade to show how the amount of sun varied, others included facial features which changed to demonstrate the changes in the sun. Some chose to illustrate temperature changes through changes in clothes, perhaps showing a woolly hat to indicate a cold day, whereas a t-shirt or sunglasses demonstrated a warmer day. Other children might select a deckchair to show a warm day while an umbrella symbolised rain. A few children represented variation in temperature by a drawing of a thermometer showing readings from "freezing cold" to "hot". A few children indicated temperature changes numerically, demonstrating an awareness of standard measures of temperature.

Figure 3.14

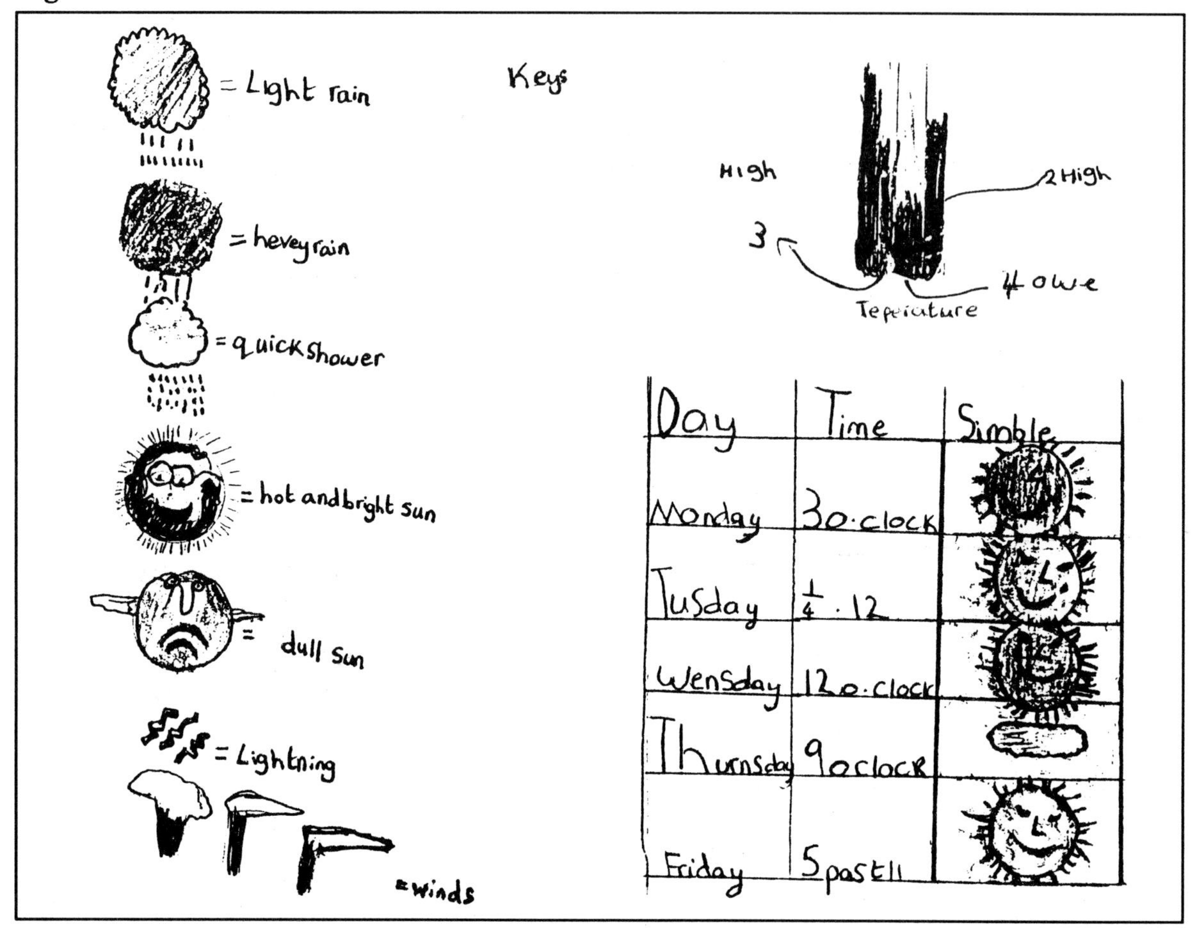

Upper junior children commonly recorded differences in windforce. They may describe a scale of windforce from "light", to "very strong". The greater effects of the wind might be illustrated by how far the branches of a tree are bent as in the Beaufort Scale. Some children also used shading to denote changes in windforce, while others invented other symbols.

3.5.2 Children's Understanding of Weather Symbols

The exploration of children's understanding of weather symbols was approached in two main ways. Some teachers asked their pupils to describe what any of the symbols they had noticed in the weather report might mean. Others prepared illustrations of the different symbols used in a weather report and asked children to describe what each symbol might mean. (Inevitably, this methodological difference affected the range of symbols described by children).

Infant children tended to describe symbols depicting the Sun, clouds and rain. Black clouds were interpreted as indicating rain. Symbols which included numbers were invariably interpreted as indicating temperature although children tended not to describe temperature in terms of degrees. A few infant children described arrows as depicting wind. However, they did not refer to windspeed or wind direction. There were no references to air pressure amongst the infants.

Upper junior children recognised a similar range of symbols to those identified by infants. They also tended to interpret symbols containing numbers as indicating temperatures rather than wind speed.

Of those junior children who indicated that the arrows depicted wind, few mentioned that the arrow illustrated the direction of the wind.

Of those children who interpreted the arrow as indicating direction some interpreted the number as indicating wind temperature.

What way the air is coming and what temperature it will be.

Few children appreciated that the arrow indicated wind direction while the number denoted windspeed.

Figure 3.15

Upper junior children rarely mentioned air pressure. Those few children who **were** aware of air pressure interpreted its effect on temperature, indicating that high pressure implied hot weather and low pressure implied cold weather.

3.5.3 Ideas about Weather Reporting

All teachers exposed their children to a television broadcast of a weather report. Children of all age groups were aware that the weather presenter was discussing the weather. The infants tended not to distinguish between the report as a **prediction** of future weather conditions and the report as a **summary** of the weather pertaining to a particular day. Junior children showed more awareness that the report was a prediction of the following day's weather.

Both infant and junior children recognised that the weather could be about local conditions or weather in other parts of the world. A few infants and an increasing number of junior children recognised that different weather conditions would prevail in different locations.

Figure 3.16

Children suggested a variety of ways in which presenters could find out about the weather. The infants tended to mention techniques which necessitated direct experience of the weather conditions. Some suggested that astronauts, or other people, would need to go out in helicopters or planes and search for direct evidence of weather conditions. Others suggested using binoculars or cameras so that weather conditions could be observed and photographed.

Spacemen go up and take photographs of the weather

While some junior children mentioned direct ways of gathering information similar to those expressed by infants, others mentioned everyday clues which people generally use to judge the weather such as the colour of the sky or the clouds.

Many junior children showed awareness of instruments which could help to describe or predict weather conditions, such as thermometers, wind dials and weather cocks. Satellites were increasingly mentioned by older children, although explanations of the ways in which these provided information varied. Some children believed the satellite contained robots which manoevered the satellite towards the Earth at regular intervals. Others described satellites as operating like cameras.

Figure 3.17

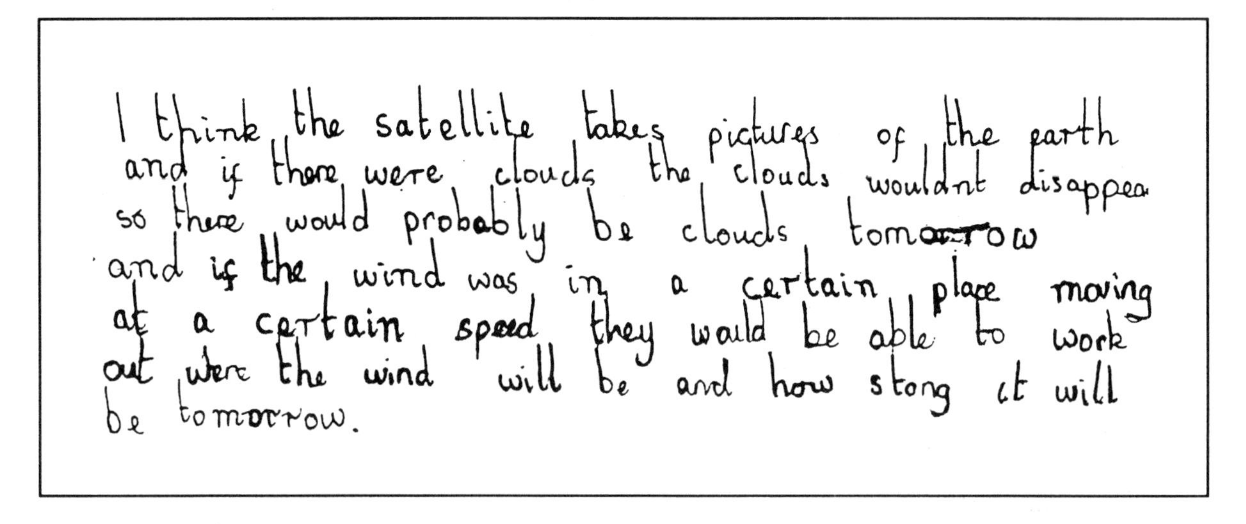
I think the satellite takes pictures of the earth
and if there were clouds the clouds wouldnt disappea
so there would probably be clouds tomorrow
and if the wind was in a certain place moving
at a certain speed they would be able to work
out were the wind will be and how stong it will
be tomorrow.

A few children were aware that the satellite transmitted signals which could be received by people on Earth.

Tune into the satellite early in the morning to find out and send signals.
The satellite provides information about clouds, wind and temperature.

A widely held view was that weather information was available from countries which experienced the weather before it was experienced locally. Children explained that people in these countries contacted the reporter to provide advance information about weather.

He will know because countries that have time ahead of us send a note about what it is like.

3.5.4 Children's Assumptions about Gender and Weather Reporting

It is tentatively suggested that children's ideas about how the weather information was gathered seemed to be influenced to some extent by the gender of the presenter. Female presenters were described as relying on other personnel, usually male, to collect weather information, whereas male presenters were considered to be actively involved in accumulating weather information.

There is a person who works with the man, he looks at the TV, he tells the man to tell the woman, who tells the weather to us.

Figure 3.18

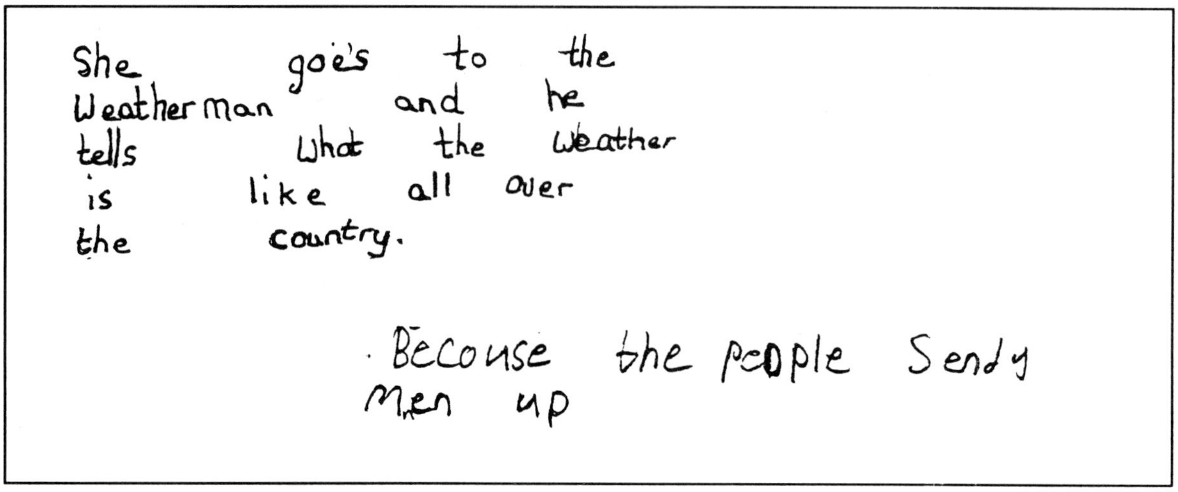

She goe's to the
Weatherman and he
tells What the Weather
is like all over
the country.

Becouse the peOple Sendy
Men up

Those children who indicated that females gathered information often suggested that women used more sedentary methods of data collection, involving less sophisticated apparatus. Male presenters were described as going out in planes to collect evidence, often using computers or satellites. The strategies adopted by males contrasted with those of females who might be described as looking out of a window for weather clues, sometimes using binoculars or waiting for information from other people or from the computer.

> *Experts go to the sea and places in other countries and they send messages to her by phone*

> *She could look out of the window to find out*

Figure 3.19

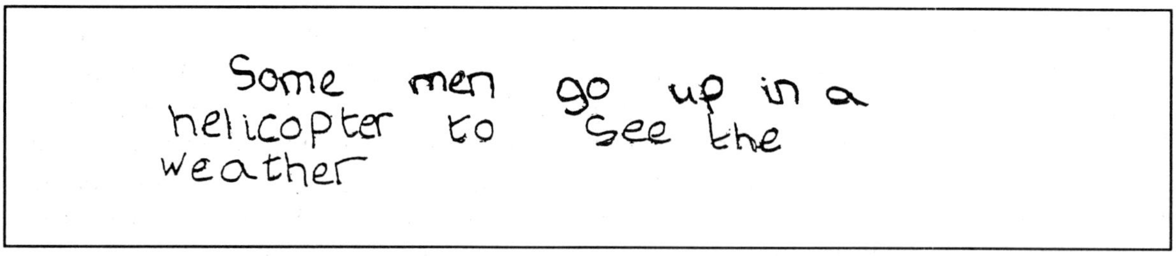

Some men go up in a
helicopter to see the
weather

3.5.5 Explaining Weather Changes

Children's attempts to explain the changes in weather patterns focused in particular on changes in rainfall and sunshine. Many children considered that clouds play a central role in weather changes. They often explained both rainy and sunny days as being determined by the clouds.

3.5.6 Explanations of Rainfall

While many children were able to explain that rain comes from clouds, their explanations of the relationship between the clouds and rain vary. Both infants and juniors tend to report that rain occurs as clouds become **heavy**. However, explanations about how the rain accumulates in the clouds were more common amongst junior children. Infants tended to suggest that the rain comes from particular clouds which hold the rain until the clouds 'burst'. This 'burst' is explained as occurring when the clouds become too heavy.

Figure 3.20 exemplifies how a child may hold an idea that particular rain clouds exist together with a notion that water from the ground is accumulated in the clouds.

Figure 3.20

Suggestions that water changes location from the ground to the clouds were rare in infant reports but more common in the accounts of lower and upper juniors. The change in location of the water was described in a number of ways. Most children considered that the water originated in the sea. Some children suggested that the water was sucked up by the clouds or that the wind gathered it up. A few children included references to changes in temperature in their explanations of how rain occurred. Some explained that water rises when it is sunny to form clouds and that as the cloud cools, rain emerges.

Figure 3.21

Although some children demonstrated an awareness of a sequence involving the transfer of water from the ground to the air and return to the ground which approximates to a water cycle, there were few references to the water changing state. Where children did indicate that water may have changed state it was referred to as "steam". (Further reference to children's understanding of the water cycle may be found in, 'Evaporation and Condensation', SPACE Report, Liverpool University Press, 1990)

Figure 3.22

3.5.7 Explanations of How Thunder Occurs

The influence of everyday expressions on children's explanations of weather changes is particularly evident as children describe how thunder occurs. Young children might suggest that "clouds are having parties", that "the angels are playing bowls", or that "God is moving his furniture".

Older children may use their knowledge of electric storms to explain the occurrence of thunder. Some believe that the clouds contain electricity which is the source of both thunder and lightening.

Figure 3.23

3.5.8 Explanations of Sunny Days

Young children may describe the moving clouds as revealing the sun. The sun may be considered to be constantly behind the clouds, revealed only when there is a gap between the clouds.

Figure 3.24

Figure 3.25

Others will suggest that the sun moves in and out, perhaps influenced by everyday expressions "the sun has gone in". Figure 3.25 has been annotated by the child's teacher, so that the idea expressed was recorded. The idea expressed is one which recurs - that temperature on the Earth is in some way connected with changes in proximity between the Earth and the Sun. When a child makes an apparently simple observation such as, "The Sun has come out", or "gone in", there are several possibilities as to the intended meaning. (Baxter, 1989).

While many children focused their explanations on rain and sunny conditions, others, particularly the upper juniors, attempted a more general explanation of changes in the weather. Older children may use their knowledge of the rotating earth to explain changes in the weather. They explain that fixed weather patterns prevail at different points and that as the Earth rotates different weather patterns are experienced.

Figure 3.26

3.5.9 *How are People Affected by the Weather?*

Children commonly mentioned ways in which people might be affected by changes in the weather. Infant children tended to focus on short term effects, including changes in clothing or references to how weather might affect their play.

Figure 3.27

Figure 3.28

Some children mentioned that their mood was influenced by weather and a few reported physiological effects such as feeling thirsty, catching cold or getting sunburned. Others mentioned how weather changes affect personal safety, pointing out that people can get trapped in snow, get blinded by the sun or drown in rough seas.

Junior children shared some of the beliefs expressed by infants, referring to how changes in weather patterns affect clothing, mood changes, leisure activities and health. Some children mentioned longer term effects of the weather on people, the vulnerability of people to weather changes, and long term effects on health, such as as frost bite or starvation.

> *I would make it rain in Ethiopia so that they could grow crops and feed themselves and then no one would die of starvation*

3.5.10 How is the Environment Affected by the Weather?

Although children were aware of some of the ways the weather affected the environment they rarely mentioned instances of weathering. Some infants focused on the aesthetic effects of weather, pointing out that rain makes the world prettier because it enables plants to grow. Young children commonly reported that the rain changed soil or sand into mud. Weathering effects which were mentioned were restricted to the effects of weather on houses.

Figure 3.29

Junior children made more frequent references to the effects of extreme weather conditions. Many children mentioned the effects of flooding indicating that damns would overflow, houses would be flooded, wildlife would be drowned and crops wouldn't grow. Others reported the long term effects of extreme high temperatures. Many suggested that plants would not grow, that rivers and seas would dry up. A few indicated that the snow would melt in polar regions bringing about flooding. There was also a recognition amongst junior children that particular weather conditions prevail in different locations. Children demonstrated an awareness that snow was prevalent in the Polar regions. They were also aware that some regions of the world, particularly Africa, were suffering from the effects of extreme high temperatures and lack of rainfall.

Figure 3.30

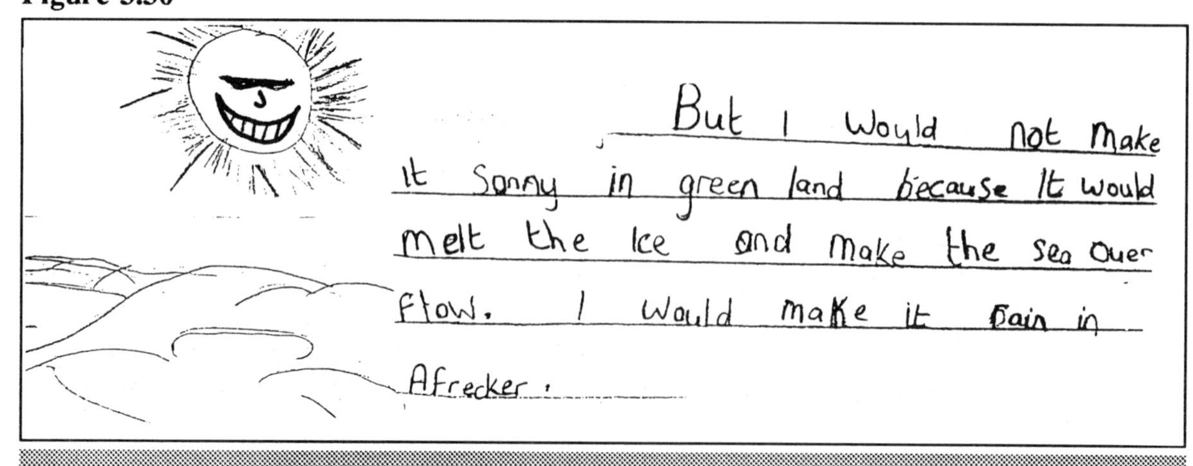

4. INTERVENTION

4.1 Intervention to Develop Ideas on Soil

The intervention described here fell into two parts. The first part extended the activity on describing soil which had been used during the exploration of ideas. The second part followed up the exploration activity of comparing soils for plant growth.

Children had been asked to say how they would find out which of two soil samples is better for plants to grow in. Many children had interpreted this question as requiring an observational response. That is, they based their answer on how the two soil samples looked, believing that certain appearances indicated the likelihood of successful nurturing of plants. Dark rich, soil spurs plant growth. Teachers were particularly interested to see whether they could foster an experimental interpretation to the question and whether children would adopt fair tests in their approach.

4.1.1 A Closer Look at Soil

Returning to the first part of the soil intervention, children were encouraged to take a closer look at soil so that they would have the opportunity to reconsider their ideas on what soil is, what it contains and how it might have been formed. Whereas in the exploration phase they were asked to look at and describe soil samples, children were now encouraged to separate the soil into whatever component parts they could find and identify. Various suggestions for means of doing this were made by teachers during the pre-intervention meeting and these are recorded in the first section on soil in the intervention guidelines (Appendix VI).

Here is an entry from an intervention diary indicating what one teacher did as a result with one part of her infant class:

> The six Year 2 children were paired and asked to look at a soil sample once again. I suggested that they might investigate their initial ideas about the 'bits' found in the samples. They were provided with:
>
> 1. straws to blow the sample to separate the bits.
> 2. sieves to sieve the sample to separate the bits.
> 3. water to separate bits - they soon recognised light (floating) bits and heavy (sinking) bits.
> 4. a hammer to crush bits.
> 5. magnifying glasses for a closer look.

The teacher noted that the children 'tended to isolate twigs, root hairs, bark, leaf mould etc, and then said that they were "in the soil", i.e. not part of it. The tendency to regard soil as a fairly homogeneous dark brown substance is a common one. With this view, the decaying plant matter only becomes soil when it has reached a sufficient state of both colour and form to blend in with the bulk of the material. It might thus be difficult to form an association between organic origins and the resultant soil. Similarly, it might be difficult to relate soil to its origins, its source as rock. This particular teacher tried to help children make this link by letting them look at the fine particles in the soil under a binocular microscope. Here is how she reports the outcome:

> *They soon realised that the fine particles were sand or bits of stone. Michael referred back to the church building and how it was possible to scrape off 'sand' from the stone.*

These children used their own experience to become aware of inorganic origins of soil. The organic link was probably still to be made.

This is how one child subsequently reported the separation process that led to his closer look at soil:

Figure 4.1

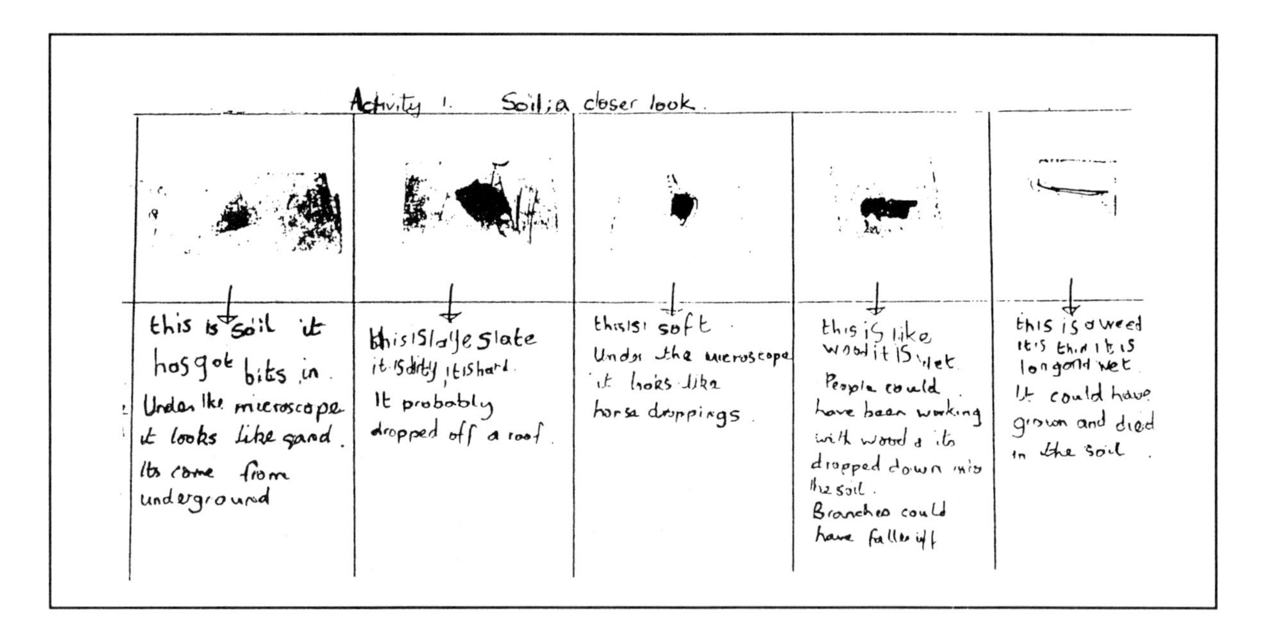

(Specimens were taped to the top of the sheet)

This child is aware of things entering the soil from above - grass stalks, slate, horse droppings, from within - dying weeds and from below - bits from underground. It is only the latter which is given the title 'soil', however, and there is no indication of any idea that the items from above and within might either be or become soil.

A group of lower junior children attempted to separate soil using various sieves such as a colander and wire mesh. Figure 4.2 indicates one child's view of the proceedings.

Figure 4.2

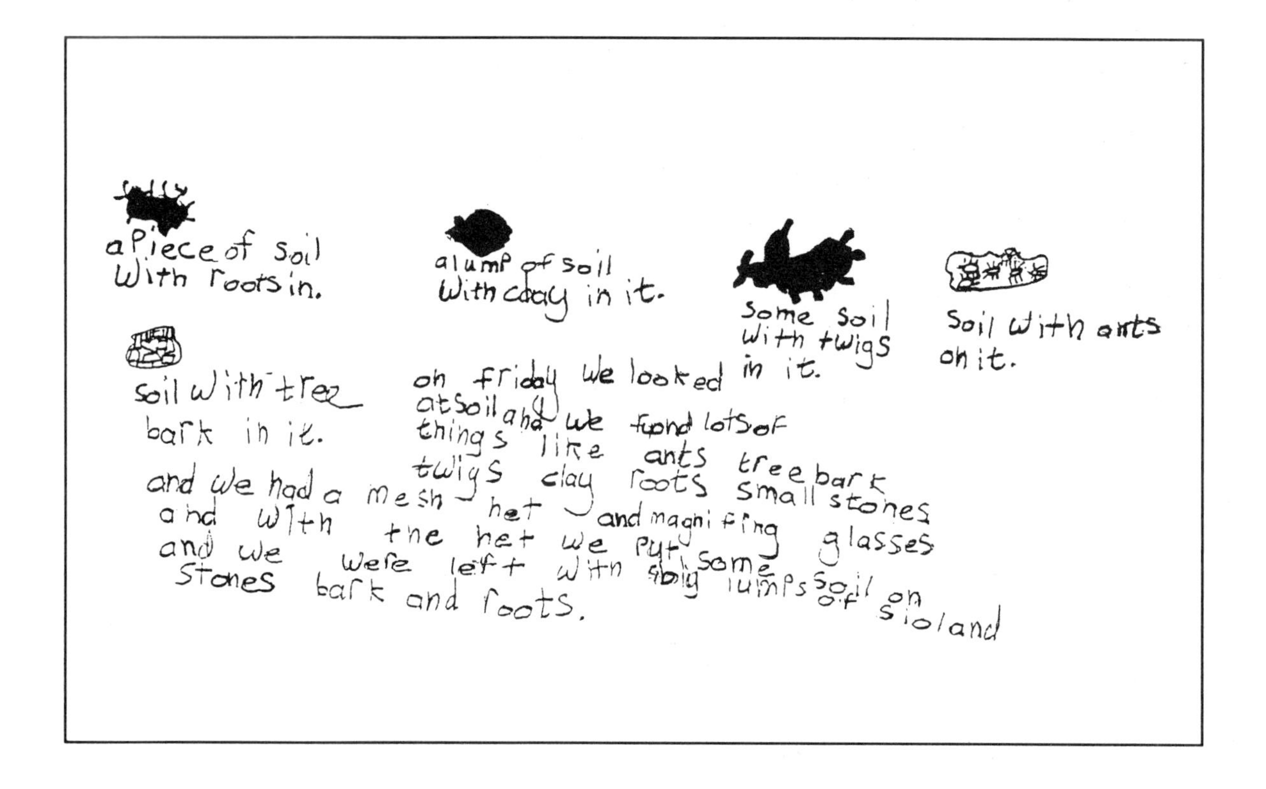

Interestingly, even clay is regarded by this child as being **in** the soil rather than being soil itself. The teacher went on to ascertain children's views on how the various components had got into the soil and this is her report of what they said:

> *The 'bits' got into the soil mainly by the wind. The clay got in the soil because of the sun. The sun changed it into a different colour. It baked it and made it harder. Roots are there because they are left behind when weeds are pulled out. Twigs are there because they've fallen off a tree or just because they've snapped.*

These children do not appear to be aware of inorganic and organic origins of soil. Instead their ideas seem to be based on personal knowledge which is likely to stem from their own experience. They may have seen soil or at least dust being blown by the wind; they may have pulled up weeds and left the roots behind; they may have seen the effect of dry, sunny weather on clay soil. Their ideas therefore reflect a view of materials as changing position (the twigs) and sometimes properties (the clay) without recognition of the possibility of one material being transformed into another.

The same group of children went on to explore what happened when stones were rubbed together. They observed a change in shape of one of the stones; they saw the powder formed. However, they made no link to the soil they had just seen. One child retained the view that the stones present in soil had got there by people throwing them in. (Figure 4.3).

Figure 4.3

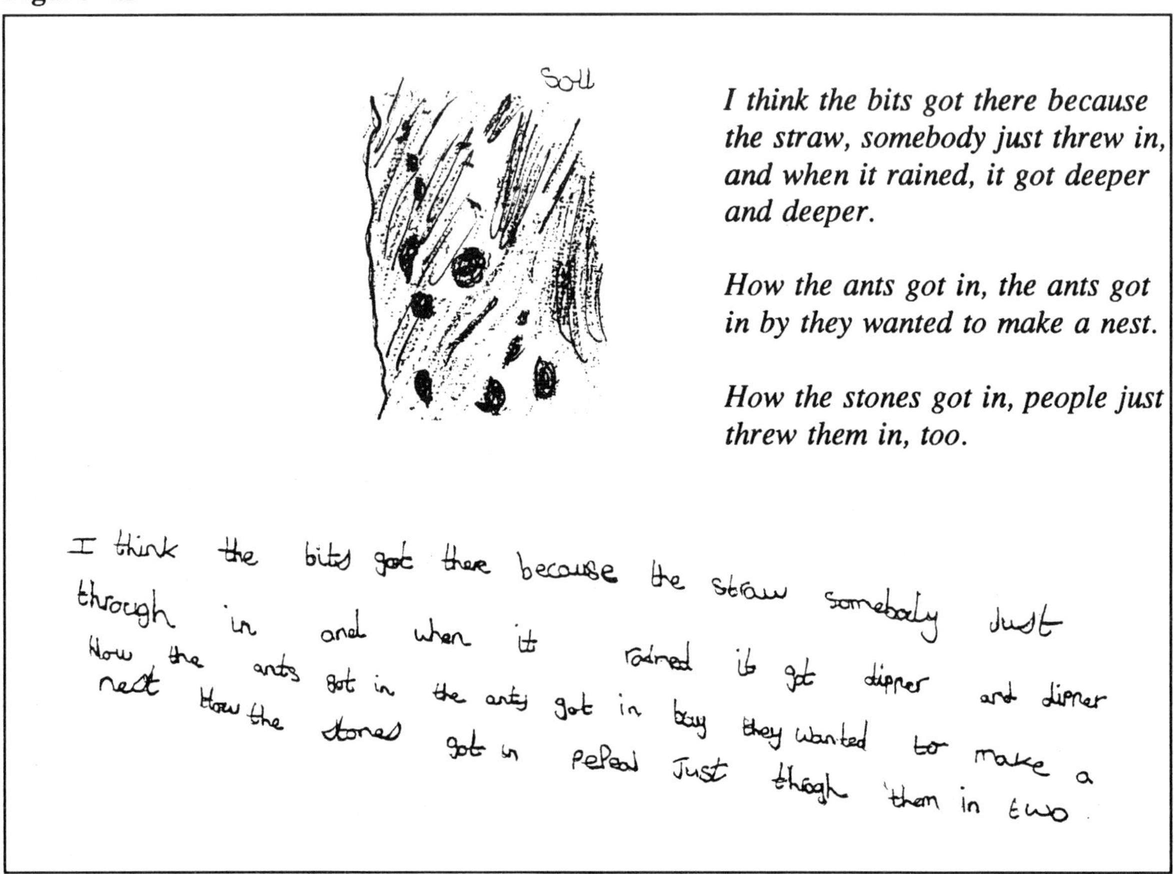

Another group of lower juniors was encouraged to separate some soil in order to find out more in relation to the ideas they had expressed during the Exploration Phase. A range of fabrics were available although the teacher did not tell the children how to use them. Very quickly, however, the children started putting soil through the fabrics. Some graded the cloth beginning with a wide mesh graduating to a very fine gauze and put each type into a separate compartment of an egg box. The teacher reported:

> *I found this uncovered a lot of questions and theories from the children. They began to see soil as a 'material' made up of a variety of substances as opposed to a substance in itself. They found it difficult to trace back the 'parts' to their origins, but had plenty of theories as to where the components had originated.*

In this class, children were working in groups for the Intervention Phase whereas earlier they had put forward ideas independently. The discussion that took place as the children shared ideas exposed them to alternative viewpoints and was an opportunity for them to reconsider their own 'theories'. Some of the children followed up the discussions by consulting books to see what ideas were propounded there. They were able to approach these books in a more critical manner because of the initial exposure to a variety of ideas, including their own.

Here is one of the ideas that had been expressed about the origins of soil (Figure 4.4.). The child who drew this sequence of pictures had clearly linked the stony component to the formation of soil even though the stones are still regarded as distinct from soil.

Figure 4.4

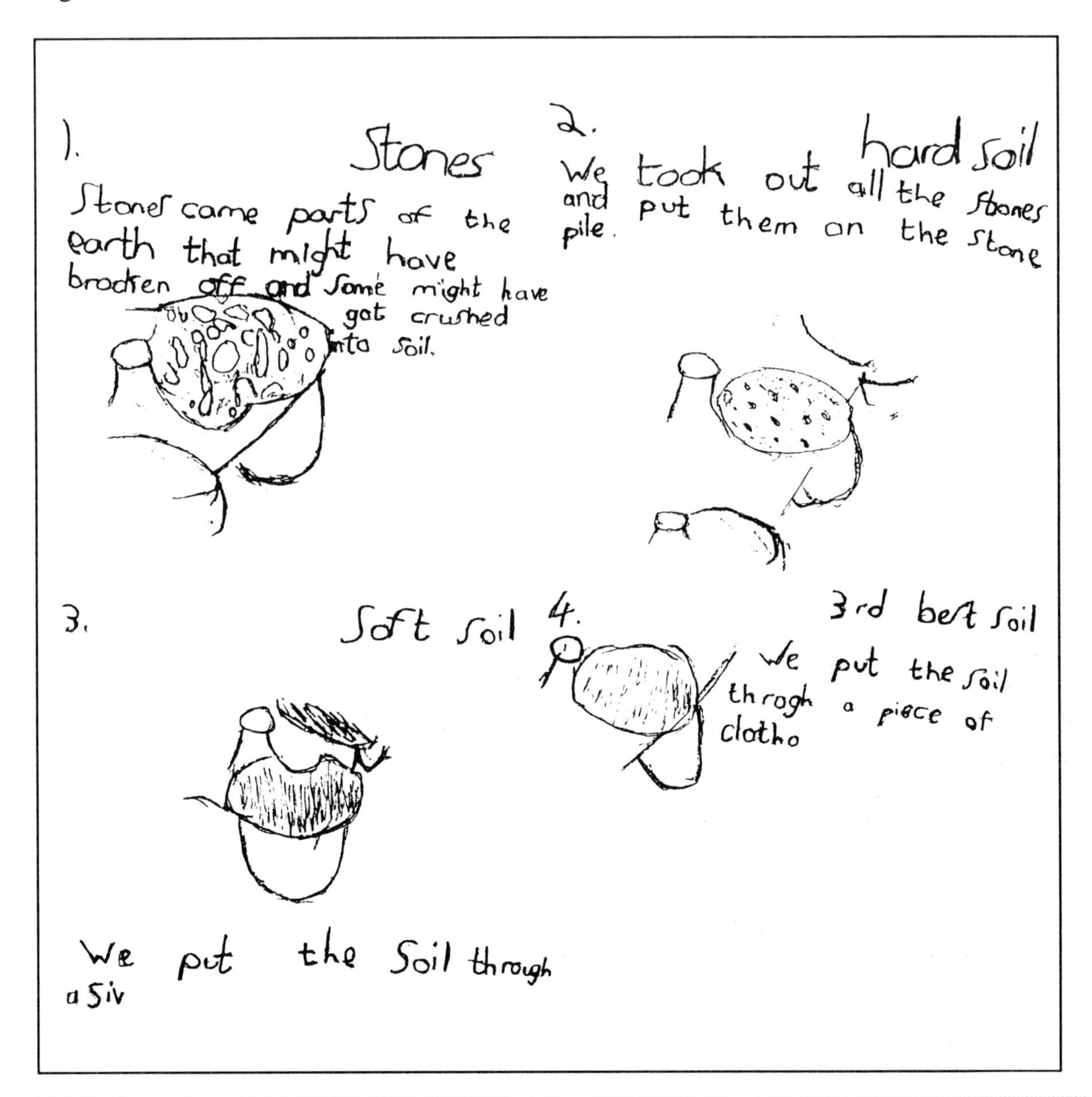

With an upper junior class, the teacher asked the children to make a list of all the things they could remember as being components of soil from the Exploration Phase. The class list was random in nature and the following is in terms of categories developed by the research team. (The terms 'organic' and 'inorganic' were not used by the children):

> various bits from plants e.g. twigs, leaves,
> various living things e.g. seeds, worms, woodlice, lumps of soil
> other inorganic matter e.g. white bits (=minerals), clay, stones
> wet or dry.

As a composite list, this is fairly comprehensive. Air is not mentioned. Children might think that there is no air present; after all you will suffocate if buried underground. Alternatively, they might recognise its presence but not consider it as part of the soil. That is, the brown bits are the soil and not the space between the bits. Again, some of the inorganic matter seems to be regarded as separate from soil. This might be because soil is expected to be brown and the 'minerals' and 'clay' are not. It also seems that the organic component is seen as something that is added to the soil rather than something that arises naturally. These children were from a farming area where they would be used to the addition of manure to the soil.

After the formation of this list, the teacher then left the children to decide on methods of separation. The techniques they suggested included using their hands, using tweezers, sieving and adding water. In the latter case they were able to make predictions about what would sink and what would float.

Another class obtained these results on adding water to soil and letting it settle. (Figure 4.5).

Figure 4.5

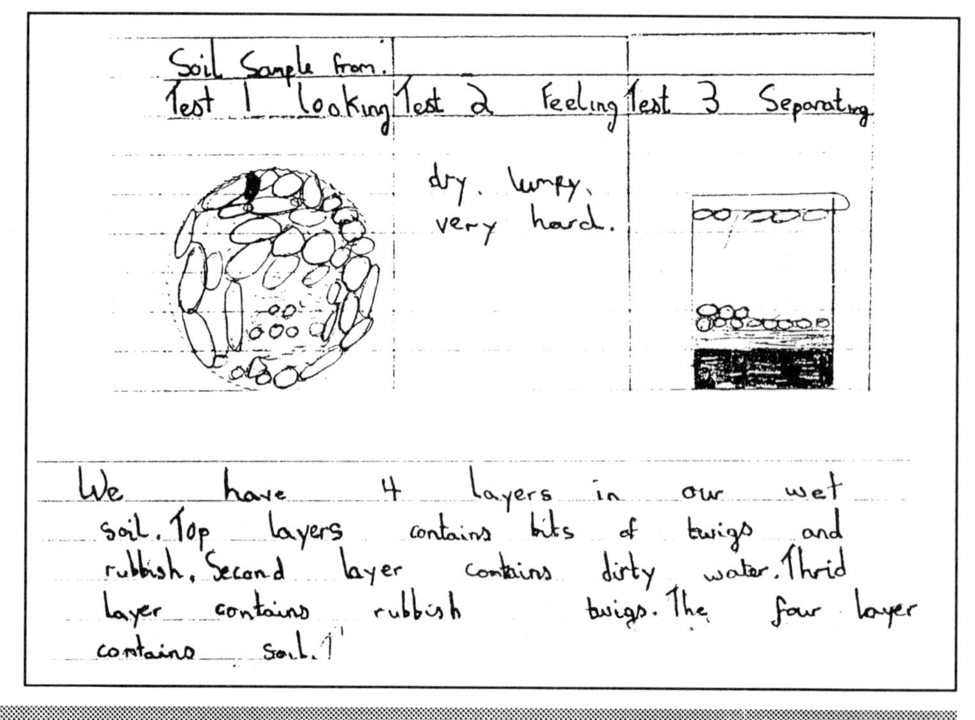

Although water had not been suggested as a universal component of the soil, the children had mentioned that soil could be wet or dry. They also had ideas about how they could detect whether there was any water in the soil. Among the suggestions in this case were, squeezing it, baking it and weighing before and afterwards and 'boiling' it and observing to see if any water vapour came off.

Even for some of these older children the origin of the lumps of soil appears unproblematic. This is shown in the following list made by an upper junior child (Figure 4.6).

Figure. 4.6

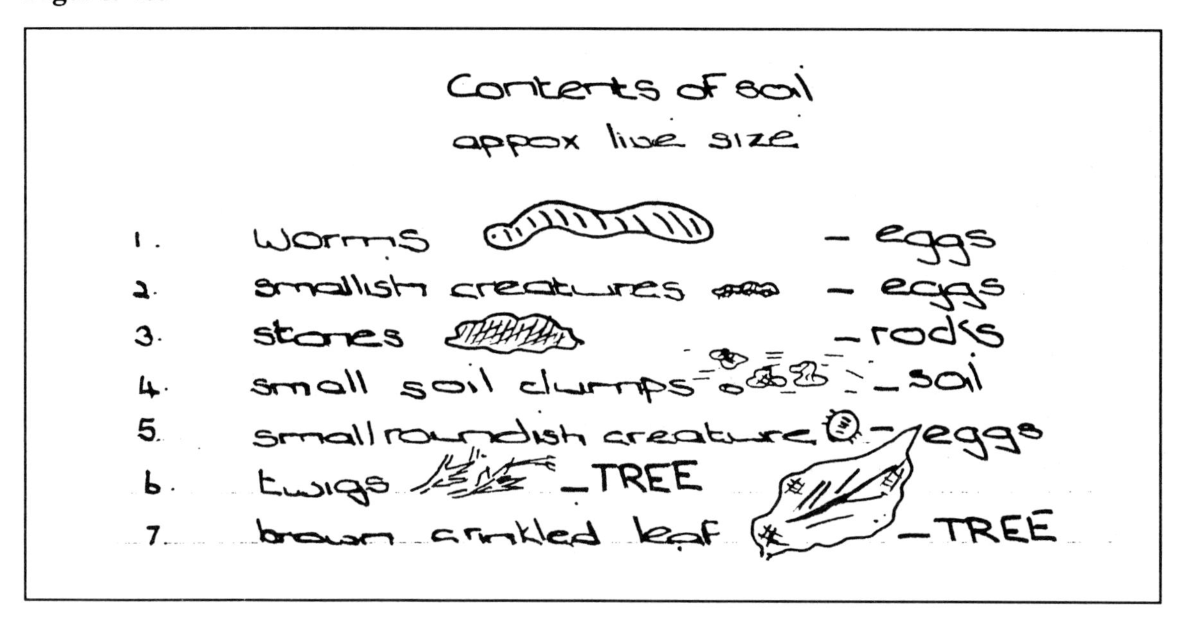

Thus the soil clumps are said to come from soil which is not traced back further. Discerning the origins of soil is difficult for many children. The inorganic processes take place over very long periods of time, are hidden and cannot be effectively simulated. Nevertheless, some children do draw on their own experience of crumbling rock and generalise to soil formation. Others recognise the varying sizes of particles in the soil from large rocks to very fine bits and are able to recognise these as various subdivisions of the same kind of matter. It is perhaps through hearing these ideas from their peers that others can be drawn into thinking about what for them has previously been unproblematic.

4.1.2 Comparing Soils

Comparing soils was the second aspect of soil intervention. The extent to which children adopted an experimental approach to the task was of particular interest. In addition, teachers explored the criteria children used in making judgements between soils to ascertain whether they had any concept of a 'good' soil.

It was clear from the exploration experiences that many children had ideas about the 'goodness' of soils based on appearance. These ideas were sufficiently strong for them to answer the question of 'finding out which soil is better' in terms of an observation. In effect, there was no need for them to investigate the soils when visual information deemed to be sufficient to decide which one was better.

The intervention was designed to see if the children could be encouraged to adopt an experimental approach to the question and determine what the nature of that approach would be. It was also hoped that children's concepts of soil differences might be expressed as they carried out the investigation and that the idea of judging on appearance alone might be seen as not always reliable.

The difference between the intervention activity and what happened during exploration was that in intervention children had to work together to make a plan and then to carry out an investigation comparing two soil samples. A number of questions designed to guide children as they plan an investigation, came out of the pre-intervention meeting and are included in the guidelines, the full details of which appear as section 2 of Appendix VI.

The following is a record of the conversation that took place between one teacher and her infant class.

T How would you find out which of these is better for plants to grow in?
C1 *Compost is the best for plants to grow in. We use it a lot at home and we have .... it's a really big plant in some.*
C2 *Ordinary soil is lumpy; compost is finer.*

Thus one child knew which 'soil' was better and the second was able to give a reason for this. After some further discussion, some children suggested a way of testing which soil was better. The conversation continued.

T How will you know which plant is growing better?
C1 *By how high it was.*
C3 *The one that grows first.*
T What will you do to make sure that you are comparing the two soils in a fair test?
C3 *Do the same things to both of them.*
C1 *Put them both in the same place and water them the same.*

It is clear from this that some young children are quite aware of how to make 'fair' comparisons. This is not always demonstrated in their writing. The following (Figure 4.7) comes from a child whose lower junior group had openly discussed the various things they should keep the same. These did not register as being the most salient factors in the investigation.

Figure 4.7

We planted some seeds
ind different soil and we loo
loocked at the soil very carefully
it had lot's of interesting things
in it and I have ejoyed doowing
it. We had to doow everything

From the same group, another child mentions some of the fairness criteria without drawing attention to them in this context (Figure 4.8).

Figure 4.8

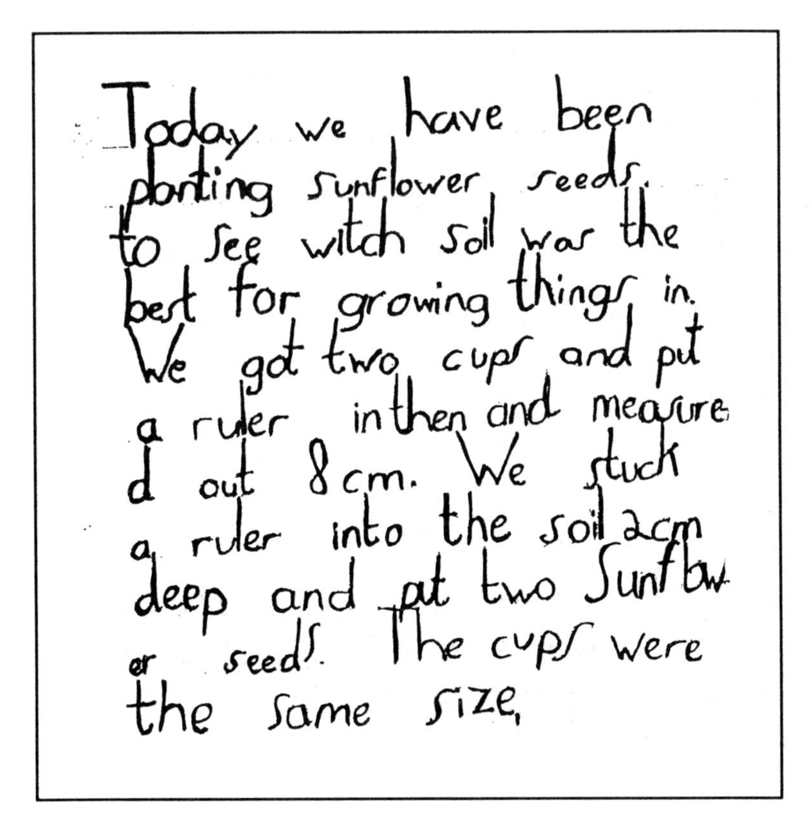

Today we have been
planting sunflower seeds.
to see witch soil was the
best for growing things in.
We got two cups and put
a ruler in then and measure
d out 8 cm. We stuck
a ruler into the soil 2cm
deep and put two Sunflow
er seeds. The cups were
the same size,

Compare what was written by the child who produced Figure 4.8 with the record shown as Figure 4.9 for a group of upper junior children. In the latter case, the use of the words, 'we will make sure' suggests that keeping things the same is much more prominent in these children's minds. They appear to be more conscious of what is important in an investigation of this nature.

Alternatively, one might say that they are becoming more aware of teacher expectations in their written accounts.

Figure 4.9

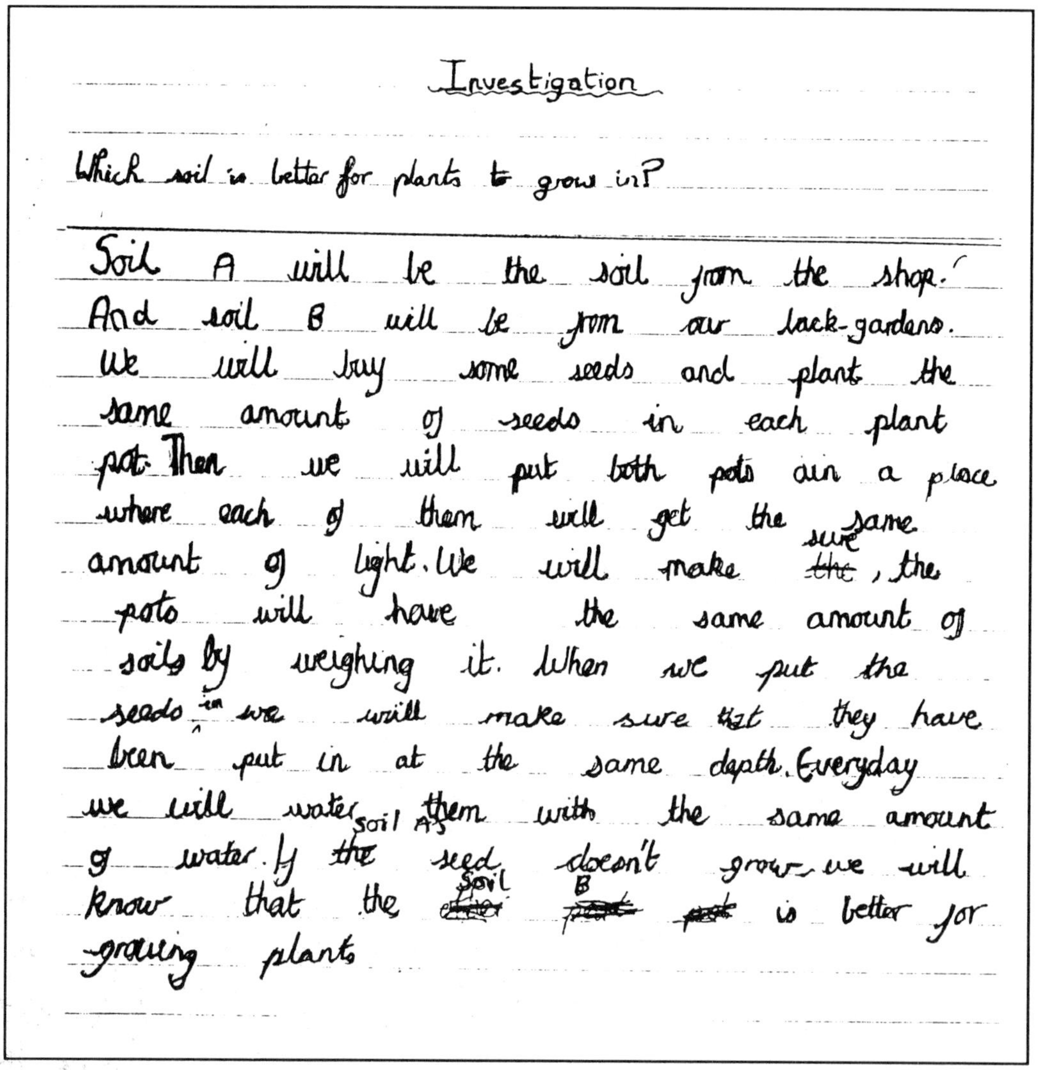

Investigation

Which soil is better for plants to grow in?

Soil A will be the soil from the shop. And soil B will be from our back-gardens. We will buy some seeds and plant the same amount of seeds in each plant pot. Then we will put both pots in a place where each of them will get the same amount of light. We will make sure ~~the~~, the pots will have the same amount of soils by weighing it. When we put the seeds in we will make sure that they have been put in at the same depth. Everyday we will water soil A them with the same amount of water. If ~~the~~ seed doesn't grow we will know that the Soil ~~other~~ B ~~plant~~ ~~pot~~ is better for growing plants

Finally, it is worth reiterating that children's personal experience has a strong influence on their beliefs. It may also affect how they think of structuring an investigation. Figure 4.10 contains the plan of a child from a small rural school. He has an idea of what colour of soil is best stemming from a knowledge of how soil is treated, and designed an investigation which appears to be a combination of an indoor control with a field trial.

Figure 4.10

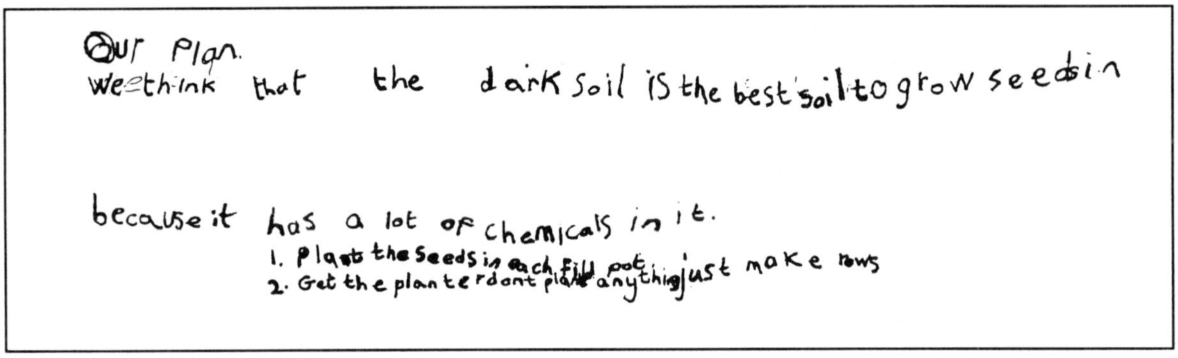

Our Plan.
We think that the dark soil is the best soil to grow seeds in because it has a lot of chemicals in it.
1. Plant the seeds in each fill pot.
2. Get the planter dont plant anything just make rows

4.2 Intervening to Develop Ideas on What is Under the Ground

The full guidelines for intervention in this concept area are given in section 4 of Appendix VI. In the Exploration Phase, children had expressed in the form of drawings their ideas about what was beneath their feet. The intention in the Intervention Phase was to ask them to consider once again those ideas in the light of any evidence they could obtain. The extent to which direct experience could be used was limited by the inaccessibility of the concept area. Teachers of the younger children, in particular, felt it was important to start with first-hand experience and did not want to introduce book knowledge at a stage when it might not be meaningful for the children. Thus one teacher commented,

> *We dug up the garden! We got to a stage of about one metre in depth and saw more and more stones mixing with the finer bits. The older infants could refer back to diagrams in a book in the classroom. I did not attempt any further work as I felt that a visit to a quarry was necessary for infant children to develop their ideas.*

Another teacher was able to provide more extended experience of what is under the ground by taking her lower junior class to see a road being dug up near the school. Not surprisingly, several of the children were more impressed by the paraphernalia of road digging than by what appeared in the hole. Moreover, the foundations of a road may not give a clear picture of what would normally exist since the original soil structure will already have been interfered with. The child who drew figure 4.11 was one of those struck by human effects.

Figure 4.11

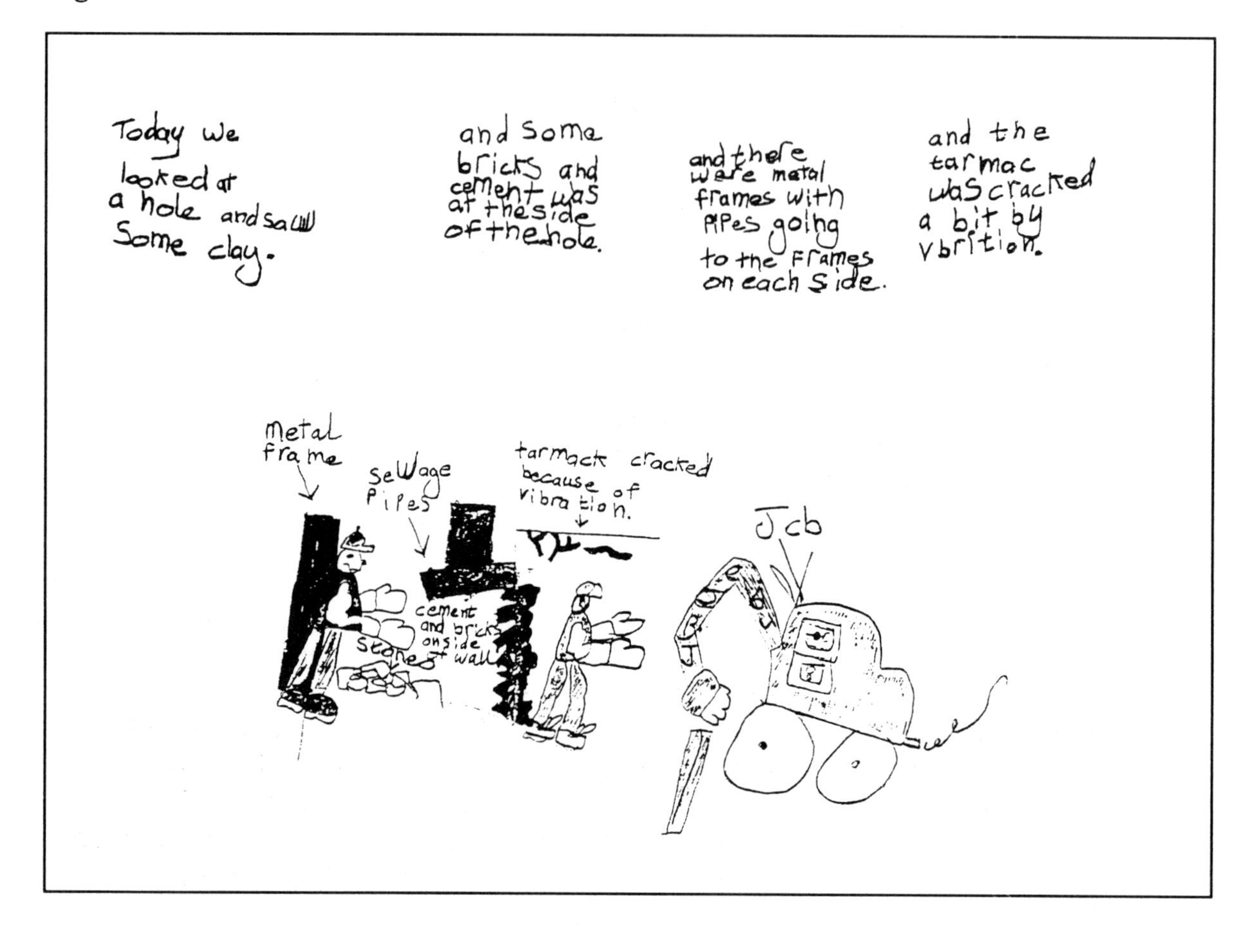

Another child made the following comments when asked whether he had seen any rock

> *I didn't see rock. I don't know if there was any. There could be further down. Rocks come from volcanoes.*

A third member of the group did see a lot of rocks and was able to establish a connection with what was underneath. He drew the picture, shown as Figure 4.12, and said,

> *The rocks have come from a rocky soil................under the Earth, it is all made of rock and mud.*

Figure 4.12

Such first hand experience can be useful in helping children realise that there is rock under the soil which itself varies in depth or may be absent altogether. There was insufficient time in intervention for most of the teachers to organise this kind of experience and so secondary sources were used. Of course, for finding out anything about the deep structure of the Earth there is no alternative to using secondary sources.

For SPACE teachers used to the approach of children testing out their own ideas, the use of books and other secondary sources of information required some accommodation. Some initially approached the use of books in a project-oriented way but realised that this often failed to make contact with children's starting points. They then asked children to recall the ideas they had expressed in the Exploration Phase and encouraged them to see whether books contained any evidence to support what they had drawn themselves. One teacher also asked her class to think about how evidence of what is underground is obtained.

Several teachers expressed the opinion that books/booklets on this topic were lacking in both variety and appropriateness. Most books contained too much factual detail and did not engage the children in considering something which is often concealed and inaccessible.

Children were also provided with opportunities to discuss their ideas. These discussions often arose in the course of using books. Despite this, the feeling remained that intervention could have been more effective had more appropriate resources been available.

4.3 Intervening to Develop Ideas on Rock

The main intention of the intervention on rock was to help children appreciate rock as a material by looking at the variety of ways in which it could be used and the variety of places in which it was found. Exploration had revealed that children were generally unaware of the widespread location of rock and that rock is under every place at which they stand.

In the pre-intervention meeting, five ways of addressing these issues were proposed and these are detailed in section 3 of Appendix VI. These five ways are now discussed under three headings. Firstly, the approaches used by teachers to develop ideas about rock in the environment will be mentioned. Then children's experiences with rock samples will be described. Finally, the use of discussion as a technique for developing ideas will be highlighted in relation to this particular topic, rock.

Teachers used a number of different approaches to try to get children to consider the various places rock was found and the ways in which it was used. One teacher of an upper junior class contextualised the activity by setting it as a problem:

> *If a builder's merchant contacted you and asked you to get as many different kinds of rock as you could, where would you go to find some and what would it look like?*

One of the group responded in these terms (figure 4.13).

Figure 4.13

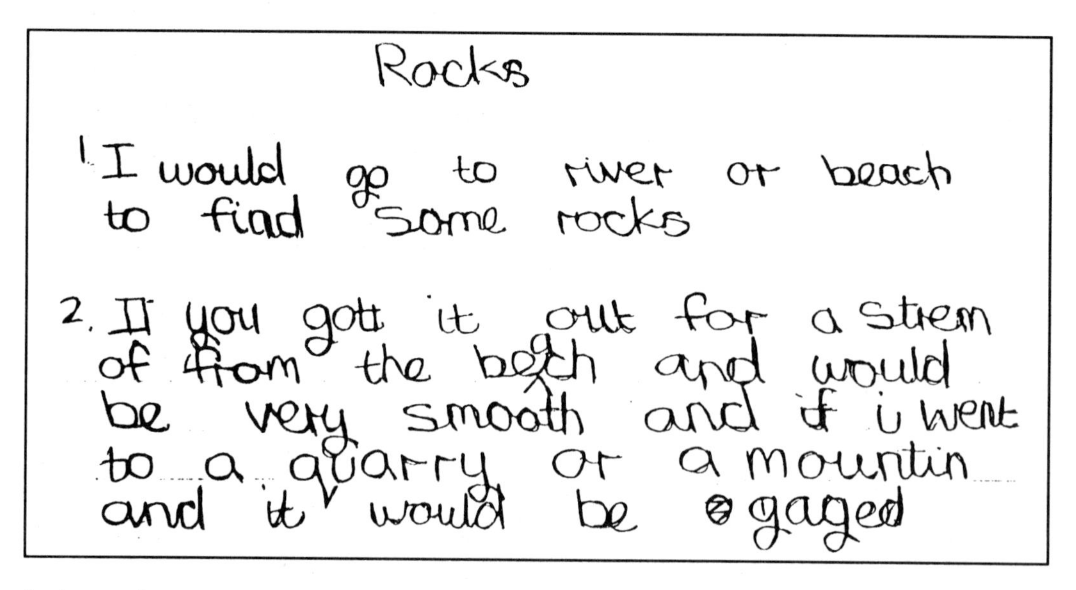

Rocks

1. I would go to river or beach to find some rocks

2. If you gott it out for a strem of from the bech and would be very smooth and if i went to a quarry or a mountin and it would be gaged

The obvious places such as rivers, beaches and mountains are mentioned. The inclusion of quarry was an opportunity to discuss where quarries are located.

Another approach to considering rock in the environment was adopted by one teacher. Her lower junior class was asked to look around the classroom and to go outdoors to make lists of items that fell into various categories. The categories decided on were metal, wood, plastic, glass and rock. The children came back and formed a composite list which for rock was playground surface, bricks, paving slabs, slates, pebbles, fence post and stone ornaments. This gave the possibility of consideration of which items had been made by people and whether some used natural materials while others used manufactured ones. An item such as 'slates' can provide an interesting point of debate if both natural slates and synthetic ones can be seen. Consideration could then be given to whether the word 'rock' could be applied to both kinds of material.

A third approach was for children to draw pictures or to collect magazine pictures and photographs and to display them. These might either be of items made from rock or of rock in its natural environment. Figure 4.14 was drawn by an infant asked to draw some pictures showing rock; the inclusion of a cave and mines provided an opportunity for discussion of whether there was always rock like that underground.

Figure 4.14

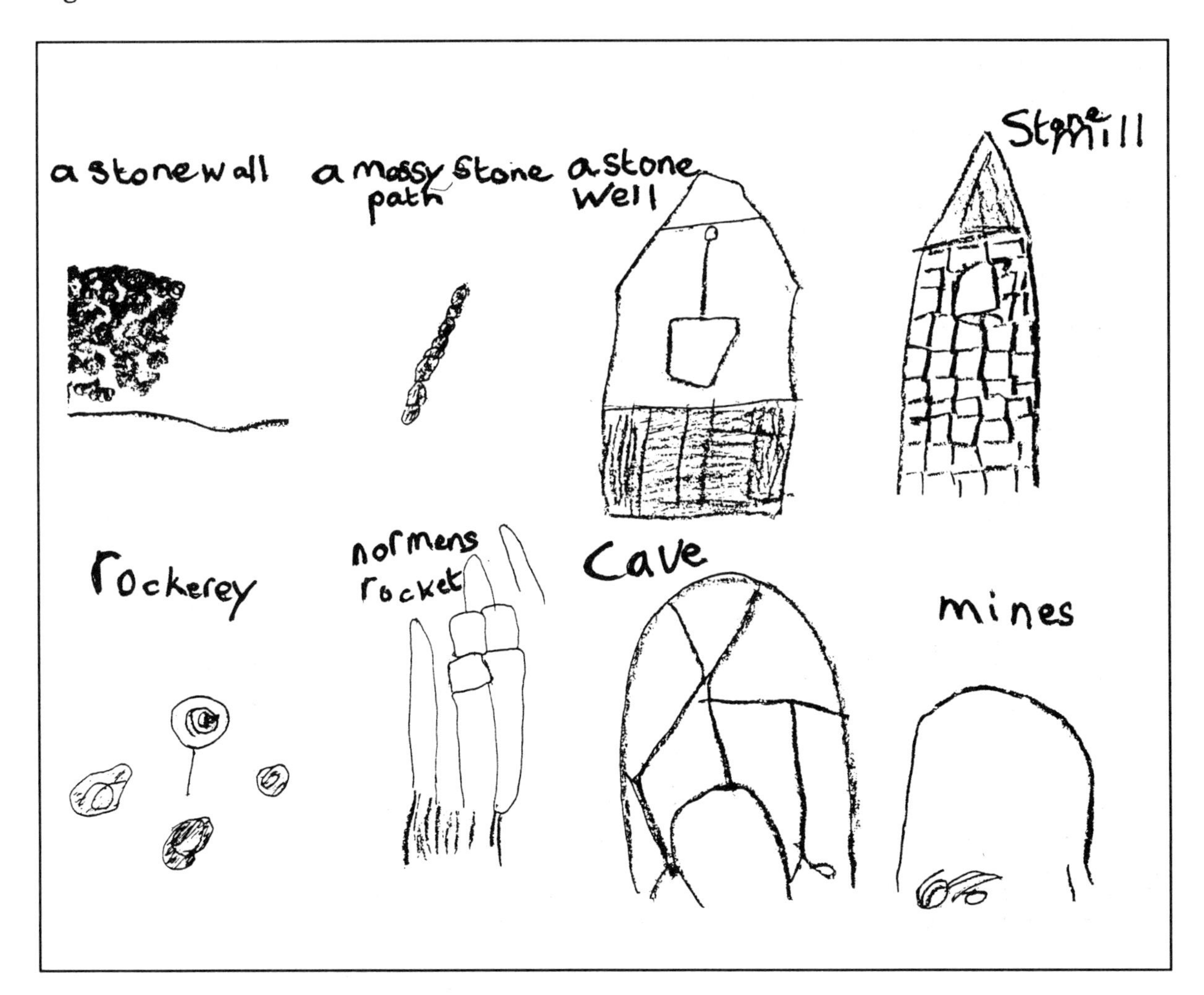

The second area of experience provided by teachers on the concept of rock involved having rock samples for children to handle, observe and discuss. With the younger children comments tended to focus on description alone. One infant, for example, looked at two lumps of rock and said,

> *One's like gold and one breaks off, the other one doesn't. One's got little cracks in it.*

Older children were more likely to give reasons for what they saw. As one teacher put it:

> *The group looked at the rock samples and attempted to group them. They were interested in them and put forward many ideas as to why they were different in colour, shape, texture and so on.*

Figure 4.15 contains drawings from three different children in that class which illustrate some of the ideas that they had.

Figure 4.15

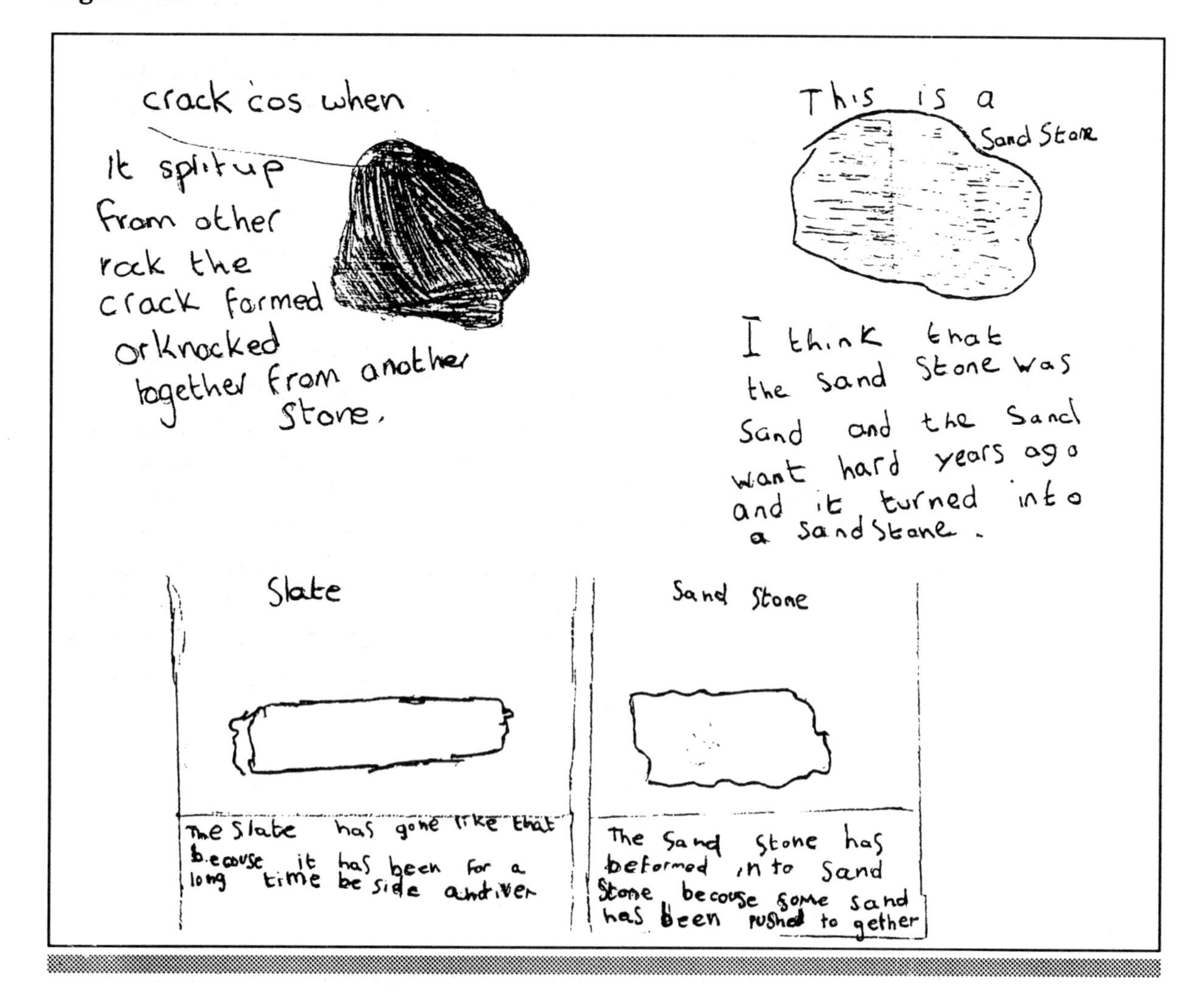

A group of upper juniors also proposed ideas about rock formation based on what they saw. The comments from one pair were:

> *We think that this coarse sandstone was formed in this shape by the wind, rain and sea because it could have been a big mass of rock and the sand, mud and moss could have been mixed with the rock and formed this coarse sandstone. The sand has also been mixed with this other kind of light sand as though it is some kind of glass or marble.*

These children suggest that there was already some rock with which the sand mixed to form sandstone rock. Perhaps they are reluctant to accept that materials as physically different as sand and sandstone can be the same material and they introduce the rock as an element of hardness. Nevertheless, sand and sandstone are probably good starting points when thinking about rock formation. The name gives a clue to the sameness of the material: the rock is crumbly and some children make a link to a reverse process of consolidation of granular sand.

The problem arose once again of how children could judge the validity of their ideas about processes as inaccessible as rock formation. Some were content if their idea seemed reasonable enough. But teachers were aware of the need to provoke reconsideration in intervention and consequently asked children to justify their ideas. Secondary sources proved useful although as one teacher commented:

> *We used the reference library at school and found some information which supported some ideas but we really needed more extensive and simply phrased books.*

Another means of provoking reconsideration of ideas was through discussion and a number of teachers used this as an intervention technique. One of the barriers to understanding rock formation and rock breaking up is the apparent hardness of rock. This, for some children, seems to give rock an aura of permanence which makes it impossible for them to contemplate any notion of creation or change. One teacher tackled this problem by encouraging a discussion after children had looked at photographs of things made from rock such as buildings. They talked about how rock could be made into the shapes which were needed. Children mentioned grinding rock down to make powder; they referred to blasting rock to make it into smaller chunks; they talked about how rock could be chiselled into shape; they even noted that some of the rock samples they had seen had small amounts of fine powder next to them.

Another group discussed the actual words they had been using. The following account is the record that the teacher made of this session:

> *I asked the children to jot down what they meant by 'rock/s' and 'stone/s' and 'pebbles'. They found this very hard. We then discussed their answers. The general opinion was that rock and stone were different - they knew rocks and stones were pieces of rock or stone. Only two children thought they were the same. The others said that there were different types of rock but only one sort of stone. They said that rocks could be underground or sticking out as 'natural materials' but stone was used to build things such as bridges, walls, statues etc. As for pebbles, they were found in water and were smooth because of the water.*

Even though at this stage children had made a distinction between rock and stone that is not normally made or, at least, is an everyday distinction, the teacher did not foreclose the discussion by rushing to a conclusion prematurely. The children went away thinking and several came back to say they had changed their minds. This is an important point about this kind of discussion. There is a tendency for it to be guided towards the point that the teacher wants to make. Yet, if the discussion is left open, it is in the time afterwards that children have the opportunity to resolve their own thoughts.

Finally, it is worth noting that during discussion children have the chance to develop their own ideas using information, from other sources. This might have come from other people or from books as in the following quotes from a discussion in an upper junior class. The children had been looking at some pieces of rock.

> *Lava comes up out of the Earth, then cools and forms mountains. It keeps on getting bigger and bigger.*
>
> *Stones gather together under the sea and form mountains.*
>
> *Plates in the world's surface squeeze together and push mountains up during earthquakes.*

A final word of caution needs to be added about the impression that this chapter gives. A rich and varied intervention has been portrayed. Teachers worked in different ways to encourage children to develop their ideas. However, the actual intervention undertaken by any one teacher or experienced by any one child was clearly much more limited than the 'overall picture' that has been presented. The time period for intervention was five weeks. Where children became absorbed in one particular area for development, other areas necessarily received less attention and occasionally were not addressed at all.

5. EFFECTS OF INTERVENTION: CHANGES IN CHILDREN'S IDEAS

5.0 Introduction

In this chapter, there is a particular focus on any shifts in children's ideas which occurred during the period of the research. In reading and interpreting the messages in this chapter it is important to appreciate the thinking behind the research design. Although the programme of work with teachers was structured to have three distinct phases - pre-intervention, intervention and post-intervention - it should not be inferred that the purpose was an 'experimental' type of study. In an experimental study, pre- and post-testing is used to measure the impact of a controlled 'treatment'. In contrast, the work reported here was far more exploratory in nature. It was not the case that there were clear views about what children **should** know or understand; nor was there sufficient information available to determine specific classroom programmes of a **remedial** nature. Experimental research can be useful when a very tightly defined area of enquiry can be identified - comparison of a drug treatment with a placebo, for instance. The research into children's ideas about rocks, soil and weather embraced much broader issues. Indeed, the boundaries were deliberately not tightly defined, other than in the sense of keeping the project manageable. The pre-intervention activities were designed to 'get a handle on' the way children were tending to think in this domain. This was not an unstructured enquiry, since there is a body of scientific ideas which provided the touchstone. However, it is a form of enquiry which demands a tolerance of a lack of closure on the part of researchers and flexibility on the part of teachers. It has much of the flavour of action research in that a lot of thinking on one's feet is needed.

The intervention phase of the research maintained the same philosophy of enquiry. Although structured to some degree, teachers' initiative, imagination and classroom experience was given full rein. It was not in the spirit of the project that a set of prescriptions for classroom activities was given or expected. There were guidelines which emerged from joint discussion and reflection on the impressions gained from children about the nature of their ideas and what looked like promising directions for activity and enquiry. At the time during which the broad scope of the intervention activities was agreed, detailed analysis and interpretation of the pre-intervention data was not yet available. What **was** available was the shared knowledge and impressions derived from first hand involvement. Some issues which, with the benefit of hindsight, might have been addressed were not. Other interventions might, in the event, not have been particularly fruitful. The important point is that much was discovered. This report aims in particular to present a broad picture of the nature of children's thinking, together with some **possibilities** for action which are based on real experiences and struggles by teachers in normal classroom environments. The issue in this chapter is not so much, 'How effective was the intervention, how big the shifts in children's ideas?'. It is rather, 'What have we learned and where do we (teachers, researchers and curriculum developers) go from here?'

5.1 *The Constituents of Soil*

5.1.1 *Total Number of Constituents*

Table 5.1 summarises the number of constituents of soil which children indicated.

Table 5.1 Total Number of Constituents to Which Children Referred

	Pre-Intervention			Post-Intervention		
Number of constituents	**Inf** n=20	**LJ** n=17	**UJ** n=21	**Inf** n=20	**LJ** n=17	**UJ** n=21
7	–	–	–	–	–	5 (1)
6	10 (2)	–	–	–	6 (1)	5 (1)
5	20 (4)	5 (1)	5 (1)	–	6 (1)	14 (3)
4	20 (4)	–	14 (3)	10 (2)	12 (2)	43 (9)
3	20 (4)	24 (4)	43 (9)	45 (9)	53 (9)	24 (5)
2	25 (5)	35 (6)	29 (6)	35 (7)	6 (1)	–
1	5 (1)	29 (5)	9 (2)	10 (2)	12 (2)	–
0	–	6 (1)	–	–	6 (1)	10 (2)
Mean number	3.6	2.0	2.1	2.6	4.7	3.8

For the sample as a whole, the mean number of constituents identified pre-intervention was 2.6; post-intervention, the mean number was 3.6. Evidently, something had changed. However, when the pattern of change is examined by age group, it is apparent that the number of constituents identified by the infants actually decreased markedly while the frequencies associated with the juniors virtually doubled. In general, children's initial ideas about soil were more simple than might be expected. For example, many children regarded soil as a homogeneous material rather than a complex mixture. Post-intervention, the infants seem to be responding with a tighter numerical range of suggestions; the older children tended to offer fewer responses suggesting just one or two components in the mixture. Another point to bear in mind in interpreting these data is that as well as the acquisition of new insights, observations and knowledge, there was also a certain amount of 'un-learning' for some children.

5.1.2 Living and Organic Constituents

The incidence of ideas about the presence of organic material in soil was of interest and a distinction was made between clear expressions of material that children suggested was **alive** (Table 5.2) and other organic material (Table 5.3).

I What is soil?

R *Mud that an earthworm's been through, quite a few times.*

I What is **in** soil?

R *Lots of creatures, insects, mini-beasts.*

I What else is in soil?

R *Weeds, roots and shoots.*

I Anything else?

R *Soil. I don't know anything else.*

Y3 B H

This boy seemed not to consider inorganic materials as constituents of soil. When observing a magnified image of sandy soil, his interpretation of the quartz granules was that they were seeds - a clear example of theory-laden observation. Exploration of soil including larger rock fragments would seem to be an appropriate experience to challenge this viewpoint of soil containing **only** organic material.

Table 5.2 Suggested Constituents of Soil: Things that are Alive

	Pre-Intervention			Post-Intervention		
	Inf n=20	**LJ** n=17	**UJ** n=21	**Inf** n=20	**LJ** n=17	**UJ** n=21
Plants or roots only	20 (4)	18 (3)	52 (11)	10 (2)	24 (4)	10 (2)
Plants and small creatures	15 (3)	6 (1)	24 (5)	30 (6)	24 (4)	28 (6)
Small creatures only	10 (2)	6 (1)	_	25 (5)	24 (4)	19 (4)
Plants, seeds and small creatures	5 (1)	_	5 (1)	5 (1)	_	_
Seeds only	5 (1)	_	_	_	_	5 (1)
Seeds and small creatures	5 (1)	_	_	5 (1)	_	5 (1)
Plants and micro-organisms	_	_	5 (1)	_	_	_
Micro-organisms	_	_	_	_	6 (1)	_
No response of this kind	40 (8)	70 (12)	14 (3)	25 (5)	24 (4)	33 (7)
Total references to living plants	50 (10)	24 (4)	86 (18)	50 (10)	47 (8)	48 (10)
Total references to living creatures	35 (7)	12 (2)	33 (7)	65 (13)	53 (9)	52 (11)

Table 3.3 in Chapter Three presented data suggesting that often children's first thoughts about soil were in terms of its **function** predominantly, the major use of soil was seen to be for **growing plants**. Table 5.2 confirms the salience of living plant material in young children's thoughts about soil. However, post-intervention, there are some interesting fluctuations. Whilst the overall level of references to living plant material remains stable at 50% amongst the infants, the incidence amongst the lower juniors doubles, while from

the upper juniors it halves. A possible hypothesis which interprets these changes positively might be that older children began to think in a more focused manner about the **component parts** of soil as a mixture, rather than simply the **contiguity** of living plants and soil. It also seems to be the case that fewer children are offering living plant material as their only suggestion of soil constituents (soil perhaps being regarded as an otherwise homogeneous material). This may be an area in which the 'unlearning' alluded to in reference to Table 5.1 might have occurred.

The major shift apparent in Table 5.2 is the increase in the number of references to living creatures to be found in soil. If children spent any time at all observing, examining and turning over soil *in situ*, an awareness of the multitude of living organic forms associated with typical garden loam would soon have become manifest.

Table 5.3 reports children's ideas about organic material other than that which is alive.

Table 5.3 Suggested Constituents of Soil: the Nature of the Organic Material

	Pre-Intervention			**Post-Intervention**		
	Inf n=20	**LJ** n=17	**UJ** n=21	**Inf** n=20	**LJ** n=17	**UJ** n=21
Plant material only	40 (8)	18 (3)	57 (12)	45 (9)	47 (8)	42 (9)
Plant and animal material	_	6 (1)	24 (5)	5 (1)	12 (2)	10 (2)
Animal material only	5 (1)	12 (2)	_	_	_	_
Decayed material	10 (2)	_	_	5 (1)	_	5 (1)
Plant and decayed material	5 (1)	_	_	_	6 (1)	5 (1)
Animal and decayed material	_	_	_	_	_	5 (1)
Plant, animal and decayed material	_	_	5 (1)	_	_	_
Totals: mention of organic content	60 (12)	36 (6)	86 (18)	55 (11)	65 (11)	66 (14)

The overall pattern is that the proportion of the sample indicating organic material was exactly the same (62%) before and after intervention, but infants and upper juniors made fewer references while the middle group almost doubled its contribution.

5.1.3 Inorganic Materials

Mention of rock in the form of sand, gravel, etc. decreased by 20% amongst the infants, increased by an approximately similar proportion amongst the lower juniors and remained stable amongst the upper juniors. Mention of water and gas was sporadic and inconsistent. References to human detritus declined overall from 9% to 5% which may be evidence of a less anthropocentric, more scientific, framework being adopted.

Whereas Table 5.4 includes rock and all its derivatives within the first response category, Table 5.5 details the various forms and the incidence with which they were mentioned.

Table 5.4 Suggested Constituents of Soil: Inorganic Materials

	Pre-Intervention			Post-Intervention		
	Inf n=20	LJ n=17	UJ n=21	Inf n=20	LJ n=17	UJ n=21
Presence of rocks, stones, grit, sand, etc.	95 (19)	65 (11)	81 (17)	75 (15)	88 (15)	81 (17)
Presence of water	15 (3)	6 (1)	10 (2)	_	12 (2)	24 (5)
Presence of human detritus	10 (2)	6 (1)	10 (2)	5 (1)	_	10 (2)
Presence of air/gas	_	_	5 (1)	_	6 (1)	_

The overall mean number of naturally occurring (as contrasted with manufactured and waste) inorganic materials remained at an average of 1.2 suggestions per interviewee both before and after the intervention. This statistic masks the fact that the number of suggestions from the infants decreased slightly while those from the lower juniors increased by an equivalent proportion.

There is an impression within the general picture of stability that a shift occurred away from mentioning 'stones' towards use of the more generic term, 'rock'. Such a shift could be important, conceptually.

Table 5.5 Natural Inorganic Materials Mentioned as Being Present in Soil

	Pre-Intervention			Post-Intervention		
	Inf n=20	**LJ** n=17	**UJ** n=21	**Inf** n=20	**LJ** n=17	**UJ** n=21
Stones	95 (19)	47 (8)	76 (16)	65 (13)	59 (10)	62 (13)
Sand	30 (6)	12 (2)	29 (6)	10 (2)	24 (4)	38 (8)
Rock	5 (1)	6 (1)	5 (1)	20 (4)	29 (5)	25 (5)
Grit	5 (1)	_	10 (2)	5 (1)	_	10 (2)
Clay	_	12 (2)	5 (1)	5 (1)	6 (1)	_
Gravel	_	_	14 (3)	5 (1)	_	10 (2)
Mean number of natural inorganic materials cited	1.4	0.8	1.4	1.1	1.2	1.4

An interesting point concerning the list of materials presented in Table 5.5 is that it closely approaches a classification of rock fragments by size. This is important because geologists (or more specifically, sedimentologists) use a set of terms with a specified range of sizes to describe rock fragments. Geologists' terms which children are not using include 'silt' and 'granule'. The use of this classification is discussed further in Chapter Six.

5.1.4 Other Suggested Constituents of Soil

While in a geological sense the nature of soil is completely defined by its living, non-living organic and inorganic constituents, other ingredients suggested by children throw an interesting light on their notion of soil. These other ideas are presented in Table 5.6.

Table 5.6 Suggested Constituents of Soil: Other Ideas

	Pre-Intervention			Post-Intervention		
	Inf n=20	**LJ** n=17	**UJ** n=21	**Inf** n=20	**LJ** n=17	**UJ** n=21
Mud/dried mud	15 (3)	12 (2)	24 (5)	5 (1)	12 (2)	10 (2)
Bits of soil/soil	25 (5)	6 (1)	5 (1)	15 (3)	12 (2)	5 (1)
Bits/lumps	25 (5)	6 (1)	5 (1)	10 (2)	18 (3)	14 (3)
Dust	5 (1)	6 (1)	5 (1)	5 (1)	18 (3)	–
Goodness/energy for plants	–	12 (2)	5 (1)	–	12 (2)	–
Minerals/chemicals	–	–	5 (1)	–	–	–
Earth	–	-	5 (1)	–	–	–
Dirt	–	–	5 (1)	–	–	5 (1)
Powder	–	6 (1)	–	–	–	–
Human Constructions	–	6 (1)	–	–	–	–
Others	–	–	–	10 (2)	6 (1)	10 (2)

It was suggested in Chapter Three that the identification of soil by re-naming it in various ways was widespread and in some instances could probably be interpreted as implying a view of soil as a homogeneous material. Table 5.6 illustrates this point in reporting children's references to 'mud', 'soil', 'earth' and 'dirt'. The combined incidence of such responses fell from an overall level of 33% pre-intervention to 21%. Such a shift is consistent with an interpretation that some enhancement of understanding has taken place. Such progression would involve giving up one notion in favour of another, more complex theory or understanding rather than only being a matter of acquiring new knowledge or understanding.

Another small but conceptually important shift is indicated by the data in Table 5.7.

Table 5.7 Reference to Different Sized Particles of the Same Constituent

	Pre-Intervention			Post-Intervention		
	Inf n=20	**LJ** n=17	**UJ** n=21	**Inf** n=20	**LJ** n=17	**UJ** n=21
Reference made to different-sized particles of same constituent	20 (4)	6 (1)	5 (1)	40 (8)	24 (4)	10 (2)
No reference made	80 (16)	94 (16)	95 (20)	60 (12)	76 (13)	90 (19)

Whereas the overall level of recognition that soil contains different sized particles of the same constituent was 10% pre-intervention, almost one quarter of the sample expressed this idea during the second interviews. This seems to be important because recognition of different-sized particles establishes a link with a view of the process of soil formation as a dynamic and continuous process, like an interpretation of a still image from the film of the event. While there is evidence of this shifting recognition, the proportion of children achieving such an insight was inversely related to age both before and after intervention, though all age groups at least doubled the number of references made to different sized particles. (The lower juniors quadrupled the number of mentions).

An insight into the fact of different sized particles existing in soil would be fairly readily supported by direct exploration of magnified images of soil samples. While hand lenses were believed to be available in all schools, only one school had access to a binocular microscope, as far as is known.

While the data in Table 5.7 have implications for ideas about the origins of soil, Table 5.8 addresses the issue more directly.

Table 5.8 Ideas About the Origins of Soil

	Pre-Intervention			Post-Intervention		
	Inf n=20	**LJ** n=17	**UJ** n=21	**Inf** n=20	**LJ** n=17	**UJ** n=21
Soil translocated	80 (16)	6 (1)	53 (11)	45 (9)	18 (3)	28 (6)
Soil transformed	–	29 (5)	14 (3)	20 (4)	70 (12)	52 (11)
Soil translocated and transformed	–	53 (9)	14 (3)	–	–	–
Soil neither translocated nor transformed	15 (3)	6 (1)	14 (3)	10 (2)	–	10 (2)
Soil always been there	10 (2)	–	10 (2)	10 (2)	–	10 (2)
No response/don't know	5 (1)	6 (1)	5 (1)	25 (5)	6 (1)	10 (2)
Other response	–	–	–	–	6 (1)	–

The suggestion initiated in the above paragraph of a more widespread recognition of soil being transformed in some way is confirmed emphatically with an increase from 14% to 47% in this type of response. All age groups made large gains, but especially the lower juniors.

I Right. So where do you think soil comes from in the first place?

R *Just, like, black soil, the rocks might have eroded, I think.*

I What does that mean? 'The rocks might have eroded'?

R *Eroded means water or... wind and water damage that gets things so that it pulls things off, like cliffs. Like the cliffs in Dover, they've been eroded because it was very soft land that was hit by the...*

I So soil comes from rocks and what happens to the rocks?

R *The rocks just stay there and fall off the cliffs.*

I So the rocks fall off the cliffs and what happens then? How do they become soil?

R *They get grinded into sand and then the sand mixes with coal... it all gets eroded into soil.*

I Good. And does anything else go in to make the soil, or does it just get made like that?

R *Nothing else, just rocks and dead animals. Things that have rotted up.*

I And what does that do to the soil?

R *It gives it, like, it makes it into compost. It gives it like compost which is good for the soil.*

Y3 B M

Translocation, the predominant idea pre-intervention, actually reduced in frequency and disappeared entirely in association with ideas of transformation. On the other hand, the proportion of children who neither suggested that soil changed form nor location declined.

Since change of form and location are geologically accurate ideas, the reduction in references to translocation might be regarded as retrograde; this issue is developed in relation to Table 5.9.

Changes in the nature and whereabouts of soil are both important considerations, geologically speaking, so the decrease in the mention of changes of location might be treated as a cause for concern. One reason for its decline might have been the tendency for children to fasten onto just one kind of explanation rather than think in terms of a combination of processes. (Handling multiple interacting data in a novel content area, might rightly be regarded as exceptional in the age group under consideration). More important is a consideration of the exact nature of the change of location which was envisaged.

Table 5.9 Origins of Translocated Soil

	Pre-Intervention			Post-Intervention		
Existing soil moved from	**Inf** n=20	**LJ** n=17	**UJ** n=21	**Inf** n=20	**LJ** n=17	**UJ** n=21
Garden	15 (3)	6 (1)	9 (2)	10 (2)	–	–
Garden Centre/shops	25 (5)	–	–	5 (1)	–	–
Fields	10 (2)	–	9 (2)	5 (1)	–	–
Under sea/water	–	–	9 (2)	–	–	–
Quarry	–	–	9 (2)	–	–	–
Other places	25 (5)	41 (7)	30 (6)	20 (4)	24 (4)	19 (4)
No response of this kind	25 (5)	53 (9)	34 (7)	60 (12)	76 (13)	81 (17)

If children are not clear about what soil is, and that it has a formative history, it is unlikely that they will be any more clear about its origins in time and location. Children's experience of soil has strong connotations of function, that function being the growing of plants. Furthermore, the prevailing view tends to be anthropocentric. So soil is associated with **people growing plants**. This orientation leads to the assumptions summarised in Table 5.9, that soil is originally located in centres of horticulture and agriculture - gardens, commercial 'garden centres' and fields. Twenty-six per cent of responses were of this nature in the pre-intervention interviews, falling to 7% post-intervention. It seems likely that the intervention procedures have served to challenge a fairly widespread misapprehension. Whether such ideas have been replaced with anything more accurate or serviceable is debateable. What is evident is that the proportion of children who were unable to offer any response to this issue has greatly increased. This might be a positive outcome if the view is taken that uncertainty is preferable to a misconception held with conviction.

The 'other places' referred to in Table 5.9 included God, shops, roads and mountains.

> *No, I don't think it's always been there, probably from the mountains, or somewhere.*
>
> Y2 B L

Table 5.10 confirms the centrality of human agency in children's notions about how soil gets to be where it is, and the diminution of this kind of thinking, from 41% to 7% overall, post-intervention.

Table 5.10 Pupils' Ideas about the Agents of Translocation

	Pre-Intervention			**Post-Intervention**		
	Inf n=20	**LJ** n=17	**UJ** n=21	**Inf** n=20	**LJ** n=17	**UJ** n=21
Human	55 (11)	18 (3)	48 (10)	5 (1)	12 (2)	5 (1)
Wind	–	6 (1)	5 (1)	–	6 (1)	–
Water/sea	–	–	10 (2)	–		5 (1)
Other (agent)	–	12 (2)	5 (1)	–	6 (1)	5 (1)
Other (no specified agent)	–	24 (4)	–	40 (8)	–	10 (2)
No response in terms of translocation	45 (9)	40 (7)	32 (7)	55 (11)	76 (13)	75 (16)

Geologists might express disappointment with the modest shifts in children's ideas apparent in Table 5.10. There is no mention of glaciation as a mechanism of translocation of material, nor of the transport of sediment by streams and rivers to estuaries. From an educational perspective, it is the 'unlearning' of the previously assumed centrality of human activity which is the most significant outcome.

5.1.6 Ideas about the Nature of the Transformation of Soil

The way in which children envisage soil might change is necessarily related to their ideas about the starting points.

Table 5.11 Ideas about the Nature of Transformation of Soil

	Pre-Intervention			Post-Intervention		
	Inf n=20	**LJ** n=17	**UJ** n=21	**Inf** n=20	**LJ** n=17	**UJ** n=21
Inorganic origins	5 (1)	29 (5)	5 (1)	10 (2)	35 (6)	33 (7)
Organic origins	–	6 (1)	19 (4)	–	6 (1)	–
Ingredients mixed	–	12 (2)	5 (1)	5 (1)	6 (1)	10 (2)
Both organic and inorganic	–	–	–	–	12 (2)	–
Cosmic origins	–	6 (1)	–	–	6 (1)	–
From inside earth	–	–	–	–	6 (1)	–
Other origins	25 (5)	29 (5)	–	–	–	10 (2)
Mud drying out	–	–	–	5 (1)	–	–
No response of this kind	70 (14)	18 (3)	71 (15)	80 (16)	29 (5)	47 (10)

The mention of inorganic origins of soil increased from the pre-intervention level of 12% to 26%, with implications of an important conceptual shift.

> *When volcanos erupted all the lava turned into rocks, rocks got grinded by other rocks and they got turned into soil.*
>
> Y3 B M
>
> *It might have come from the rocks, might have grounded down and then dropped in muddy patches.*
>
> Y1 G H

On a smaller scale, there were more suggestions of soil comprising a mixture of materials in the second interviews (from 5% to 10%) including two lower juniors who specified that soil is a mixture of organic and inorganic materials.

All the stones rub together it forms a soil and then the leaves rot away it gives moisture. The water dries up but it stays in the soil.

Y3 G M

With regard to the agents of transformation, the now familiar picture of the initially widespread attribution of human agency diminishing during the course of the study is repeated.

All the sea covered the sand. As all the rain came down, all the sand came to the top because of all the rain lifting it up, and all the insects made it dark and it changed to a different colour

Y2 G H

Any change in the form or position of rocks and soil, however small, involves some form of transfer of energy. The agents of geological change involve immense forces: the sea, or more specifically, the movements of the oceans which cover two thirds of the Earth's surface under the influence of the gravitational force of the Moon; the movement of water through precipitation and flow in rivers (or 'high energy transport systems' in geologists' terminology) together with the processes of dissolving and deposition; movements of air related to temperature, density and pressure differences; the stresses of heating and cooling from within the planet and from the Sun. Children's responses showed increased awareness of these agencies. There is an interesting issue as to just how conceptually accessible these ideas are to children in the age range under consideration. It might be one thing to be informed of and to memorise a list of agents. It is quite another to see the forces operating at the level of the whole system of the Earth within the Solar System, which is the only perspective which offers any sort of coherent and comprehensive picture.

Table 5.12 summarises children's ideas about how it might be explained that soil has changed in **form**. The agents of change have been identified. Human agency was deemed to be less prevalent than was the case in expressions of ideas about how soil changed location, but a parallel diminution of this idea is apparent.

The recognition of the role of water in general and the sea in particular showed large gains. This may be because the enormous energy transfers associated with the sea meeting land forms can be readily witnessed. The increased awareness of 'collision' as an agent of change is also possible due to the perceptual immediacy of this process, as compared with other, imperceptible, erosion processes.

Table 5.12 Ideas about the Agents of Transformation

	Pre-Intervention			Post-Intervention		
	Inf n=20	**LJ** n=17	**UJ** n=21	**Inf** n=20	**LJ** n=17	**UJ** n=21
Humans	10 (2)	24 (4)	5 (1)	5 (1)	12 (2)	5 (1)
Sea	_	12 (2)	_	15 (3)	71 (12)	52 (11)
Sun	_	_	5 (1)	_	6 (1)	5 (1)
Rain	_	_	5 (1)	5 (1)	12 (2)	10 (2)
Wind	_	_	5 (1)	_	6 (1)	_
Collision	5 (1)	_	_	_	35 (6)	24 (5)
Frost/ice/glaciation	_	_	_	_	_	_
Other	5 (1)	12 (2)	19 (4)	5 (1)	12 (2)	29 (6)
No response (in terms of Transformation)	95 (19)	18 (3)	71 (15)	85 (17)	29 (5)	48 (10)
Average number of suggestions per pupil, non-human agents	0.05	0.1	0.1	0.2	1.3	0.9

5.1.7 Criteria Used in Reaching Decisions as to Whether Samples are of Soil

In order to draw out the criteria which children were operating to define a material as soil or not soil, five samples of relevant materials were presented to them. These were: sandy topsoil, sand from a sandpit, small pebbles, chalky soil and damp peat. Children were asked,

Which of these would you call soil?

Their responses were then probed with respect to the categorisation of each material by asking,

What makes you call this one soil?

or alternatively,

What makes you say that this isn't soil?

Judgements about the materials being soil or not, together with expressions of uncertainty, are summarised in Table 5.13.

Table 5.13 Judgements about Materials Being Soil or Not Soil

Sample		Pre-Intervention			Post-Intervention		
		Inf n=20	**LJ** n=17	**UJ** n=21	**Inf** n=20	**LJ** n=17	**UJ** n=21
Garden soil (sandy topsoil)	soil	80 (16)	100 (17)	86 (18)	45 (9)	82 (14)	71 (15)
	not soil	20 (4)	_	14 (3)	50 (10)	18 (3)	24 (5)
	uncertain	_	_	_	5 (1)	_	5 (1)
Sand from sandpit	soil	5 (1)	24 (4)	24 (5)	_	_	_
	not soil	95 (19)	70 (12)	76 (16)	_	_	_
	uncertain	_	6 (1)	_	_	_	_
Small pebbles	soil	5 (1)	_	5 (1)	_	_	_
	not soil	95 (19)	100 (17)	90 (19)	_	_	_
	uncertain	_	_	5 (1)	_	_	_
Chalky soil	soil	20 (4)	18 (3)	43 (9)	5 (1)	35 (6)	48 (10)
	not soil	80 (16)	76 (13)	57 (12)	85 (17)	59 (10)	48 (10)
	uncertain	_	6 (1)	_	10 (2)	6 (1)	5 (1)
Damp peat	soil	100 (20)	94 (16)	90 (19)	95 (19)	76 (13)	81 (17)
	not soil	_	6 (1)	5 (1)	_	6 (1)	19 (4)
	uncertain	_	_	5 (1)	5 (1)	18 (3)	_

As suggested in earlier sections of this chapter, a certain amount of shaking of existing beliefs as well as the acquisition of new views was encountered.

The **sandy topsoil**, though a fairly typical sample of loam from the region, was the cause of some doubt before the intervention activities and even more after.

> *Sandy. It's the same colour of soil. Soil has bits in it sometimes, that's a bit like sand.*
>
> Y4 B H

> *It's not soil. We walk on sand on the beach and it's not like mud and things.*
>
> Y3 G L

> *It's not soil. It's very small grains and I can see quite a lot of sand, but I can't see very much else.*
>
> Y6 G M

The **sand** was of the type which children would be likely to have encountered at school or in domestic sand-pits, being fine grained, and yellow-brown in colour. While 17% had viewed this sand as an instance of soil pre-intervention, none did so during the second round of interviews.

The **small pebbles** produced a result similar to that for the sand though fewer children had regarded this material as an instance of soil initially.

The **chalky soil** caused more uncertainty post-intervention than it had previously. Children tended to reject the chalky soil as an instance of soil on the basis of colour. Many suggested that it was a manufactured material such as cement or concrete.

> *No, it's , I think it might be cement, something like cement, I think.*
>
> Y3 B M

> *Cos it looks like concrete mixing and it's very light in colour.*
>
> Y3 B L

> *It's concrete, kind of. More stone than soil, like concrete.*
>
> Y3 G M

> *It's chalk. You know when you make cement, it's that stuff you put in to mix cement. It makes your hands white and might burn you.*
>
> Y1 B M

The **damp peat** provoked slightly more rejection from the upper juniors and slightly less acceptance by the lower juniors. In this sample in particular, the criteria which children were using to make their judgements are important. If they are operating a definition of soil as having a high inorganic content, then it is likely that peat will be rejected as an instance of soil.

It is easy to sympathise with children's dilemmas in deciding whether or not to regard peat as an instance of soil.

Table 5.14 Descriptions of the Peat Sample

	Pre-Intervention			Post-Intervention		
	Inf n=20	LJ n=17	UJ n=21	Inf n=20	LJ n=17	UJ n=21
It is soil, it's peat	–	18 (3)	5 (1)	15 (3)	29 (5)	29 (6)
It is soil, it's compost	5 (1)	6 (1)	–	5 (1)	18 (3)	14 (3)
It's compost, but uncertain about soil	–	–	–	–	6 (0)	–
It is not soil, it's compost	–	6 (1)	–	–	–	14 (3)
It is not soil, it's peat	–	–	5 (1)	–	6 (1)	5 (1)
No reference to peat or compost	95 (19)	70 (12)	90 (19)	80 (16)	41 (7)	38 (8)

Peat is a spongy material derived from plant material, especially mosses, which have partially de-composed in a wet environment such as a bog. Dead plant material decomposes in anaerobic conditions and becomes compacted over the years. Children's definitional problems are probably compounded by their tendency to perceive soil as a growing medium for plants and the appropriation of the terms 'compost' and 'potting compost' by horticulturalists. Whereas the etymology of 'compost' is based on manure or mould (as in leaf-mould, decomposed vegetable matter), the range of proprietary 'potting composts' certainly include those with a proportion of inorganic material, typically sharp sand which improves the drainage qualities of the material.

Almost a third of the lower and upper juniors (24% of the sample as a whole) treated the peat as an instance of soil, or perhaps even more simply, as a re-naming as had been the case with 'mud', 'dirt', 'earth' and so on.

About nine per cent of the sample explicitly viewed soil and compost as equivalents and therefore agreed that the peat was an instance of soil.

It might be hoped that one of the effects of the intervention activities would be that children were more analytical in their observations of soil-like materials. The suggestion was made in Chapter Three that there was some tendency for children's observations of samples to be limited to surface features. For example, great significance might be attached to the property of colour while features such as texture and composition might be neglected or ignored. Given the range of soil colours between relatively white and black, with the addition of reds and ochres, colour is unlikely to be a very powerful classifying characteristic. Table 5.14 records the frequency of reference to colour.

Table 5.15 Extent of Recognition of Soil by Reference to Colour

	Pre-Intervention			**Post-Intervention**		
	Inf n=20	**LJ** n=17	**UJ** n=21	**Inf** n=20	**LJ** n=17	**UJ** n=21
Colour only mentioned	30 (6)	29 (5)	24 (5)	35 (7)	18 (3)	5 (1)
Colour and another reason(s) mentioned	30 (6)	47 (8)	19 (4)	25 (5)	29 (5)	24 (5)
Other reason(s), excluding colour, mentioned	35 (7)	18 (3)	52 (11)	40 (8)	53 (9)	66 (14)
No response/don't know	5 (1)	6 (1)	5 (1)	–	–	5 (1)

Reference to colour both as the sole deciding criterion and in combination with other attributes declined amongst the two oldest groups. The incidence of reasons for deciding that a material was soil by reference to qualities other than colour increased post intervention and was also more frequent with age.

		Pre intervention.	Post intervention
A	R	*It's sand.*	*It's a sandy soil.*
	I	Is sand not the same as soil?	*It's got sand in it and little pebbles.*
	R	*No, it's a different colour.*	
D	R	*It's not a soil. It's a different colour.*	*It's not soil. It's in big lumps.*
	I	What colour does soil have to be?	
	R	*Brown or black.*	
E	R	*It's soil because it looks like soil.*	*The peat is soil. It's got roots in it.*

Y6 B L

(A = sandy topsoil; D = Chalky soil; E = peat)

5.2 Ideas about Rocks

5.2.1 Sand as an Instance of Rock

In general, at the beginning of the study, there seemed to be evidence of a tendency for children to regards stones, pebbles, sand and rock as identifiable but separate entities rather than materials sharing the attributes of a common parent material. In posing the question of the relationship of sand to rock, the intention was to explore ideas about the continuity of a material in fragments of different sizes. A sample of beach sand was presented and children were asked whether they thought the sand was rock. Responses are summarised in Table 5.16.

Table 5.16 Thoughts on Whether 'Beach Sand' is Rock

			Post-Intervention	
		Infants n=20	**Lower Juniors** n=17	**Upper Juniors** n=21
Unqualified	No	60 (12)	47 (8)	38 (8)
Unqualified	Yes	25 (5)	47 (8)	52 (11)
Qualified	Yes	10 (2)	6 (1)	10 (2)
Don't know		5 (1)	–	–

While the majority of the youngest age group rejected the proposition that the sand was an example of rock, the lower juniors were split down the middle in their judgement, while slightly more than half of the upper juniors responded positively. Since inappropriately superficial judgements such as size had been revealed to be criterial in classifying material as rock during the initial interviews, this could represent an important shift. (This question had not been posed, initially, since its diagnostic value only emerged later. Consequently, there is no initial response frequency against which to make comparisons).

> *No, I call it sand. Sand is softer than rock, rock is harder than sand. Rock is like stones.*
>
> Y1 B M

Table 5.17 summarises the reasons which children presented for accepting sand as an example of rock.

Table 5.17 Reasons for Accepting Sand as Rock

	Post-Intervention		
	Infants n=20	**Lower Juniors** n=17	**Upper Juniors** n=21
Sand derived from rock	10 (2)	36 (6)	48 (10)
Sand may become rock	10 (2)	–	–
Share rock and sand property of hardness	–	6 (1)	5 (1)
Other reasons	15 (3)	12 (2)	5 (1)
Sand not accepted as rock	60 (12)	47 (8)	38 (8)
Don't know	5 (1)	–	5 (1)

In geological terms, sand is loose rock or mineral material resulting from mechanical processes such as disintegration or abrasion and moved from its place of origin. Such a precise definition and understanding was not expected of children. The responses which are grouped together in the first row of Table 5.17 are those in which children seemed to indicate the realisation that sand was a derivative of larger masses of rock and was indeed the same substance.

Two children, both infants, also showed awareness of the fact that sand may be reconstituted as rock.

> *Some of it is rock. The shells at the bottom press on to the sand and mud sticks to the shell. It gets hard and turns to rock. Everything's protecting it and it's protected too much and it's squeezed together and turns to rock. Rock is like hard and when you bang it it comes out as sand. So rock is made of sand.*

Y2 B H

> *Yes, it is rock, because when it gets washed up it goes into rocks. When the seawater slams onto the little rocks the water sticks to rocks and gradually the little rocks stick together, and build into a very big rock.*

Y2 G H

It may be that the word 'sandstone' coupled with the granular texture of this particular kind of rock make a direct and intuitive impression even on young children. If this is the case, it would seem likely that the older children have not been exposed to this idea or were aware but did not judge the information relevant to the discussion.

Children's arguments against the case of classifying the sand as rock are summarised in Table 5.18.

Table 5.18 Reasons for not Accepting Sand as Rock

	Post-Intervention		
	Infants n=20	**Lower Juniors** n=17	**Upper Juniors** n=21
Has a different size	15 (3)	12 (2)	14 (3)
Has different properties (hardness)	15 (3)	24 (4)	5 (1)
Has a different name	10 (2)	–	5 (1)
Doesn't look like rock	10 (2)	–	–
Has different properties (colour)	–	6 (1)	5 (1)
Different location	10 (2)	–	–
Not interconvertible	–	–	5 (1)
Other reasons	–	6 (1)	5 (1)
Sand is rock	35 (7)	53 (9)	62 (13)
Don't know	5 (1)	–	–

The most frequently cited argument against sand being rock was the most salient perceptual difference, that of size, though only 8 (58%) indicated this as their determining criterion. At the same frequency level, the property of hardness/softness was mentioned. It seems highly likely that children have in mind the difference between falling on or being struck by rock as compared with sand when they make this assertion. (A similar difficulty with use of the term 'hardness' is discussed in the SPACE Research Report, 'Materials', Section 3.12 (Russell et al, Liverpool University Press, 1990)).

Other reasons included differences in name, appearance, location, colour and a denial of the possibility of one material being capable of being transformed to the other.

> *No, because sand looks light brown and the rocks look all different colours. They're hard. They're different shapes, they're called different names. Some are marbly colour. Some can be oval and some round.*
>
> Y3 G H

5.2.2 A Rounded Sample as an Instance of Rock

According to the geologist's definition, the rounded sample of sandstone, being in the range 64-264 mm would be designated a 'cobble'. The sample was presented to children and they were asked whether it was rock. Responses are presented in Table 5.19.

Table 5.19 Categorisation of Rounded Sample as Rock or Not Rock

		Post-Intervention	
	Infants n=20	Lower Juniors n=17	Upper Juniors n=21
Unqualified Yes	80 (16)	41 (7)	71 (15)
Unqualified No	15 (3)	29 (5)	19 (4)
Uncertain	–	29 (5)	10 (2)
Missing data	5 (1)	–	–

The infants and upper juniors each produced a majority response which accepted the sample as rock. The response of the lower juniors did not fit the pattern; they were both more rejecting of the sample as rock and also expressed more uncertainty in their judgements. Table 5.20 sets out the main reasons which swayed children's decisions. Note that many children offered more than one reason in support of their decision.

Table 5.20 Reasons for Categorisation of Rounded Sample as Rock or Not Rock

	Post-Intervention		
	Infants n=20	Lower Juniors n=17	Upper Juniors n=21
Hard/unbreakable, so rock	50 (10)	6 (1)	29 (6)
Smooth, so rock	30 (6)	24 (4)	19 (4)
It's a 'stone'/'pebble', so not rock	15 (3)	29 (5)	19 (4)
It looks like rock	20 (4)	6 (1)	19 (4)
Large, so rock	20 (4)	_	19 (4)
It's a 'stone'/'pebble', so it's rock	5 (1)	18 (3)	14 (3)
Heavy, so rock	20 (4)	6 (1)	5 (1)
Small, so not rock	5 (1)	12 (2)	14 (3)
Smooth, so not rock	_	18 (3)	14 (3)
Colour indicates it's rock	5 (1)	6 (1)	14 (3)
Soft, so not rock	_	12 (2)	_
It doesn't look like rock	_	12 (2)	_
Light, so not rock	_	6 (1)	_
Colour indicates it's not rock	_	6 (1)	_
It feels like rock	-	6 (1)	-
It doesn't feel like rock	-	6 (1)	-
Another reason, so rock	5 (1)	12 (2)	24 (5)
Another reason, so not rock	_	12 (2)	_
It's (rock name mentioned) so rock	_	6 (1)	_

The lower juniors seemed to be more susceptible to the distraction of vocabulary; more than the other two age groups they suggested that the rounded sample was not rock because it was a 'pebble' or a 'stone'. They also referred to smoothness and the fact of the sample being 'soft' as disqualifying the sample from being a rock.

The positive responses refer to the material's hardness - an attribute which would be of interest to the geologist. Smoothness also figures significantly, though decreasingly with age. The range of surface characteristics are in evidence as they were in the first round of interviews. One child offered a more specific name for the rock material.

What is it realistic to expect of this age group in terms of their analysis of materials? This is a difficult question to address, one which must be approached empirically, rather than *a priori*. What children seemed **not** to do was to consider the **material** of which the sample was made. Their judgements seemed to be more to do with classifying an object than thinking more analytically about the material of which that object is comprised. The decision seemed to be treated as 'all or nothing': 'rock is rock ', rather than thinking about the material's hardness, texture, constituent components, porosity, presence or absence of crystals, and so on. It could be that the children comprising the interview sample, following a brief exploratory period during which teachers also were finding out the limits and possibilities for supporting understanding, lacked sufficient experience of the right kind. For example, it would seem that children need to consider a whole range of common but distinguishable rock forms and explore their similarities and differences along a number of dimensions. Initially, these could be the methods which children invent, discover or learn from one another. There is also a case for some input about the techniques and conceptual frameworks which geologists use, providing these are intellectually accessible, safe and practicable.

There would also seem to be scope for the development of appropriate technical vocabulary which might serve the purpose of sharpening children's perceptions, again with the proviso that the vocabulary must be closely linked to checks on conceptual understanding.

5.2.3 Rock and Soil

Children's ideas about the range and continuity of the inorganic material of the world were further explored by seeking their views about possible relationships between rock and soil. Table 5.21 presents their ideas about the possibility of rock being transformed into soil.

Table 5.21 Thoughts on the Possibility of Rock Transforming to Soil

	Post-Intervention		
	Infants n=20	**Lower Juniors** n=17	**Upper Juniors** n=21
No, rock cannot change to soil	60 (12)	35 (6)	38 (8)
Yes, rock can change into soil	30 (6)	35 (6)	43 (9)
Maybe rock can change into soil	5 (1)	29 (5)	10 (2)
Some can, some can't	5 (1)	–	–
No response	–	–	10 (2)

Very approximately, two thirds of the infants did not entertain the possibility of rock being capable of changing into soil; a similar view was elicited from a little over one third of the lower and upper juniors. Overall, 36% confirmed the view that rock could change to soil, this trend increasing slightly with age.

There is a fine point to be considered which is whether children were doubtful (as a number were, especially amongst the lower juniors) because the soil would not comprise inorganic material alone. Only two children made an explicit comment on the need for the addition of organic materials for the transformed rock to be considered to be soil, while another two responses suggested the need for other material to be added without indicating what this should be.

> *Yes, when the rock gets rubbed against walls and things. It needs all the other things to become soil - sand and leaves fall down, the leaves rot away slowly and go into the soil and that helps the soil to produce a soil. It's not really made without leaves.*
>
> Y3 G M

The more precise reasons which children adduced in support of their judgements are summarised in Table 5.22.

Table 5.22 Reasons for Denying the Possibility of Rock Transforming to Soil

	Post-Intervention		
	Infants n=20	**Lower Juniors** n=17	**Upper Juniors** n=21
Rock is hard and soil is soft	25 (5)	24 (4)	14 (3)
Rock becomes sand	15 (3)	–	14 (3)
Wrong colour	–	12 (2)	10 (2)
A rock is always a rock	10 (2)	6 (1)	–
Rocks are solid	–	–	14 (3)
Rock is not fine enough	–	–	5 (1)
Don't know/no response	15 (3)	–	5 (1)

The property of hardness of rock is the attribute which most pupils tend to hold on to and which blocks recognition of the possibility of rock contributing to soil. As was the case with the comparison with sand, children see soil as being 'soft' and as such, lacking an important defining property of rock. Assumptions about the solidity, size and colour of rock also barred acceptance of rock as convertible to soil. Three children gave responses of the 'rock is rock' variety, implying that the intervention activities had not succeeded in supporting the re-structuring of the basic idea that 'rock' is not a unidimensional entity but a material to be found in a whole range of guises.

The reasons given by those children who did accept some form of transformation of rock to soil are summarised in Table 5.23.

Table 5.23 Reasons for Believing that Rock may be Transformed into Soil

	Post-Intervention		
	Infants n=20	**Lower Juniors** n=17	**Upper Juniors** n=21
Rock crumbles, breaks down	_	_	48 (10)
People bang and crush rocks	5 (1)	24 (4)	10 (2)
Rocks broken down by water/rain	10 (2)	6 (1)	14 (3)
Personal experience of grinding rock	10 (2)	6 (1)	_
Rocks broken down by other stones	_	18 (3)	_
Rocks broken down by machines	_	6 (1)	5 (1)
Rocks broken as time passes	_	6 (1)	_
Rocks broken down by heat	5 (1)	_	_
Rocks broken down by people walking	_	6 (1)	_
Rocks broken down by machines (as only cause)	_	_	5 (1)
Rocks broken down by little creatures	_	6 (1)	_
Rocks broken down by magic	5 (1)	_	_

The essential realisation that children must grasp in order to accept the possibility of rock being convertible to soil is that the size, shape and form of rock is a variable over time. The range of responses which were encountered all made sense, geologically speaking, though there were omissions of processes which are significant - volcanic and tectonic effects, glaciation, etc. The realisation that rocks can be broken down at all is progress. There is some evidence that the effects are attributed to people breaking up rocks, using machines, walking over them, or other means. There is also reference to the passage of time and other acceptances of outcomes without specification of causes. The effects of water and/or rain were mentioned by 10% of the sample.

5.2.4 The Concept of Continuous Rock Underground

No clear age-related pattern emerged with respect to the idea of continuous rock existing underground. (See Table 5.24).

Table 5.24 Acceptance of Continuous Rock Underground

	Post-Intervention		
	Infants n=20	**Lower Juniors** n=17	**Upper Juniors** n=21
Yes, accepted	45 (9)	18 (3)	67 (14)
No, not accepted	35 (7)	71 (12)	5 (1)
Uncertain	5 (1)	12 (2)	29 (6)
Don't know	15 (3)	–	–

Two thirds of the upper juniors showed allegiance to this idea while slightly more than two thirds of the lower juniors did not. About one third of the infants rejected the notion while just under half supported it. Additionally, there were expressions of uncertainty rather than rejection of the notion by most of the upper juniors who did not expressly support the notion of continuous rock underground.

The reasons for rejecting the idea of continuous rock are reported in Table 5.25.

Table 5.25 Reasons for not Accepting Continuous Rock Underground

	Post-Intervention		
	Infants n=20	**Lower Juniors** n=17	**Upper Juniors** n=21
Rock can't stretch that far	20 (4)	35 (6)	5 (1)
No reason for response	5 (1)	6 (1)	24 (5)
Rock not underground	10 (2)	–	–
It would break up	–	12 (2)	–
No continuous rock, lots of small rocks	–	12 (2)	–
It would take too long to put together	–	6 (1)	–
Other response	–	12 (2)	5 (1)

The overriding impression from the negative expressions is of children struggling conceptually with the scale implied by the question. The idea of a rock as discrete entities seems to preclude the logical possibility of it existing on such a whole-world scale. Suggestions that such an expanse of rock could not hold together, but would be bound to break apart, are the result.

> *You couldn't get one that long. Not even mountains could be that long.*
>
> Y3 B H

> *No, it couldn't go as far as Blackpool, underground. Something might break. There will be rocks all the way to Blackpool.*
>
> Y3 B M

Similar doubts about the possibility of rock existing on such an enormous scale are apparent in the positive responses, also, as revealed by the data presented in Table 5.26.

Table 5.26 Further Details of 'Continuous Rock Underground' Responses

	Post-Intervention		
	Infants n=20	**Lower Juniors** n=17	**Upper Juniors** n=21
Continuous rock, not necessarily ubiquitous	_	_	33 (7)
No reason for response	5 (1)	6 (1)	14 (3)
Knowledge of rock under earth	5 (1)	_	10 (2)
Plain acceptance	5 (1)	6 (1)	5 (1)
Rock all joined together	_	6 (1)	10 (2)
Omnipresence of rock	10 (2)	_	_
Omnipresence of rock but maybe joined together	_	_	10 (2)
Other reasons	10 (2)	_	10 (2)

One third of the responses of the upper juniors suggested an agreement with the idea of rock being continuous underground. However, even with probing, it could not be firmly established that these children held the belief that such continuous rock was ubiquitous, throughout the planet.

> *Yes, it would be made up of different rocks, all joined together. They've been stuck together for years and years.*
>
> Y5 G H

> *If you made a hole all the way through like the Channel Tunnel, they've dug through rocks to get through. Yes, there is rock all the way through.*
>
> Y5 B H

Another sizeable group (19% of the sample) did not present any further elaboration in support of their stated belief. Others indicated that they were simply reporting a known fact (knowledge of rock under the earth) or simply accepted the idea with no further consideration.

5.3 Ideas about What is Below the Surface of the Earth

5.3.1 Ideas about Layers

Children's ideas about form and structure below the Earth's surface were explored through the medium of drawings as described in Chapter Three. In this section, changes in the representations of the sample are described and discussed. No significant differences were found in the salient characteristics of drawings starting from tarmac or grass, so these are treated as equivalent.

How far children decided to take their drawings - whether to the centre of the Earth, or beyond, or as a complete section of the planet - was a matter of personal choice. There was a small increase from 2% to 9% in the proportion of children whose drawings included the centre of the Earth, though no such response was collected from the infants. Table 5.27 shows the ways in which the cross-sectional drawings illustrated layers, if at all.

Table 5.27 Ideas About Rock Layers

	Pre-Intervention			**Post-Intervention**		
	Inf n=20	**LJ** n=17	**UJ** n=21	**Inf** n=20	**LJ** n=17	**UJ** n=21
Continuous rock layer	10 (2)	6 (1)	33 (7)	_	24 (4)	52 (11)
Separate rocks within layer	15 (3)	18 (3)	19 (4)	15 (3)	18 (3)	5 (1)
Rocks joined within layer	_	_	19 (4)	_	_	24 (5)
No mention of rock layers	75 (15)	76 (13)	29 (6)	85 (17)	59 (10)	19 (4)

Amongst the juniors, there was a marked increase in the proportion of children who showed **rock layers** in the internal structure of the planet. About one half of the upper juniors and a quarter of the lower juniors indicated a **continuous rock layer** under the surface. There was a decrease in the number of children showing separate rocks within these layers.

This particular aspect of understanding of the Earth's structure is not one which is amenable to an experiential approach. The techniques for exploring the Earth's structure with seismic impulses involve concepts and sequences of reasoning which would be very difficult to make accessible to children other than in a very simplified sense.

Consequently, even the evidence which geologists make use of is not really interpretable by children in the elementary phase. For these reasons, any knowledge which children have about the structure of the Earth is likely to originate from secondary sources (as contrasted with primary sources of direct experience). This may be the reason for the shifts in representation towards more accurate ideas being associated with the older children in the sample. It may be that teachers were more likely to regard the use of secondary sources as appropriate to the older age groups. If it was the case that the exposure was undifferentiated by age, it might be the case that such information is more easily understood, remembered and recalled by older children. Teachers reported difficulties in locating materials which they regarded as suitable.

The more general nature of the layers (including layers other than of rock) as represented by children are discussed by reference to Table 5.28.

Table 5.28 General Nature of Layers Drawn

	Pre-Intervention			Post-Intervention		
	Inf n=20	LJ n=17	UJ n=21	Inf n=20	LJ n=17	UJ n=21
Layers of finely divided particles and rock layers	15 (3)	_	24 (5)	15 (3)	12 (2)	62 (13)
Layers of finely divided particles and technically named layers and rock layers	5 (1)	6 (1)	24 (5)	_	35 (6)	19 (4)
Layers of finely divided particles only	25 (5)	_	24 (5)	10 (2)	6 (1)	10 (2)
Layers of finely divided particles and technically named layers	_	18 (3)	28 (6)	_	6 (1)	5 (1)
Alternatively named layers	5 (1)	23 (4)	_	_	6 (1)	5 (1)
Unnamed layers	_	6 (1)	_	_	_	_
Drawing did not contain layers	50 (10)	47 (8)	_	75 (15)	36 (6)	_

All the drawings produced by the upper juniors, both before and after intervention, displayed a layer formation. The proportion of infants drawing layers increased from 50% to 75%, while the lower juniors recorded a slightly decreased incidence.

The most commonly occurring representation of layers included layers of finely divided particles, fragments or granules, together with layers of rock. (See Figure 5.1). Almost two thirds of the upper juniors recorded this model, far more than was the case with the other two age groups.

The second largest response group was where the lower juniors showed the greatest increase post-intervention. This was the category in which layers of rock, particles and technically named layers were included.

Figure 5.1 Typical Layer Drawing

The particular content of the layers was varied, as Table 5.29 affirms.

Table 5.29 Specific Content of Layers Drawn

	Pre-Intervention			Post-Intervention		
	Inf n=20	**LJ** n=17	**UJ** n=21	**Inf** n=20	**LJ** n=17	**UJ** n=21
Includes soil	30 (6)	18 (3)	57 (12)	20 (4)	59 (10)	57 (12)
Includes water	30 (6)	6 (1)	38 (8)	15 (3)	_	43 (9)
Includes clay	5 (1)	29 (5)	29 (6)	_	24 (4)	38 (8)
Includes both clay and sand	10 (2)	_	38 (8)	10 (2)	18 (3)	14 (3)
Includes mud	20 (4)	_	24 (5)	_	_	10 (2)
Includes sand	10 (2)	6 (1)	24 (5)	5 (1)	12 (2)	14 (3)
Includes tiles / bricks	_	12 (2)	19 (4)	5 (1)	6 (1)	_
Includes named rock types	_	_	19 (4)	_	_	5 (1)
Includes layer of 'rock'	10 (2)	18 (3)	33 (7)	15 (3)	12 (2)	71 (15)
Includes named rock types and a layer of rock	5 (1)	_	10 (2)	_	_	5 (1)
Includes both soil and mud (as separate layers)	_	6 (1)	10 (2)	5 (1)	6 (1)	14 (3)
Includes oil	_	_	14 (3)	_	6 (1)	10 (2)
Includes coal	_	_	14 (3)	_	12 (2)	5 (1)
Includes oil and coal	_	_	_	_	_	10 (2)
Other layer types	5 (1)	24 (4)	48 (10)	5 (1)	24 (4)	43 (9)
No response in terms of layers	50 (10)	47 (8)	_	75 (15)	35 (6)	_

Notable post-intervention shifts include increases in inclusion of soil by lower juniors; a decrease in reference to manufactured tiles and bricks from 10% to 4% amongst all juniors; an increase in inclusion of a layer of rock by the upper juniors.

The number of layers children presented in their drawings was strongly related to age. (See Table 5.30).

Table 5.30 Number of Discreet Layers Shown

	Pre-intervention			Post-intervention		
	Inf n=20	**LJ** n=17	**UJ** n=21	**Inf** n=20	**LJ** n=17	**UJ** n=21
8 Layers	_	12 (2)	19 (4)	_	6 (1)	19 (4)
7 Layers	_	_	10 (2)	10 (2)	_	10 (2)
6 Layers	_	6 (1)	33 (7)	_	12 (2)	14 (3)
5 Layers	_	_	10 (2)	_	18 (3)	29 (6)
4 Layers	5 (1)	29 (5)	24 (5)	5 (1)	12 (2)	10 (2)
3 Layers	15 (3)	_	5 (1)	5 (1)	18 (3)	14 (3)
2 Layers	25 (5)	6 (1)	_	5 (1)	_	5 (1)
1 Layer	55 (11)	47 (8)	_	75 (15)	35 (6)	_
Mean number of layers per pupil	1.2	2.5	5.8	1.9	3.4	5.4

The use of technical terms to describe the internal structure of the Earth was not in evidence amongst the infants but occurred sporadically amongst the juniors. (Table 5.31)

Table 5.31 Use of Technical Terms to Describe the Internal Structure of the Earth

	Pre-Intervention			Post-Intervention		
	Inf n=20	**LJ** n=17	**UJ** n=21	**Inf** n=20	**LJ** n=17	**UJ** n=21
Mentions 'core' only	–	41 (7)	43 (9)	–	24 (4)	24 (5)
Mentions 'mantle' only	–	–	–	–	–	5 (1)
Mentions 'core and mantle'	–	12 (2)	–	–	12 (2)	–
Mentions 'core and crust'	–	6 (1)	–	–	–	5 (1)
Mentions 'crust' only	–	–	–	–	6 (1)	5 (1)
No technical term used	100 (20)	41 (7)	57 (12)	100 (20)	59 (7)	62 (12)

The use of the term 'core' was the most frequently encountered technical term, but even this decreased from 33% to 21%. The term 'crust' was used by one child pre-intervention and three during the follow-up interviews. Two children used the word 'mantle', increasing to three post-intervention.

5.3.2 Drawings Without Layers

Rarely in this kind of research is the empirical data as unequivocal as that presented in Table 5.32: the 'no layer' drawings were obtained **only** from the younger children.

Table 5.32 Contents of 'No Layer' Drawings

	Pre-Intervention			Post-Intervention		
	Inf n=20	**LJ** n=17	**UJ** n=21	**Inf** n=20	**LJ** n=17	**UJ** n=21
Living creatures	40 (8)	24 (4)	_	55 (11)	12 (2)	_
Animal/plant remains	20 (4)	18 (3)	_	15 (3)	18 (3)	_
Pipes	25 (5)	12 (2)	_	10 (2)	12 (2)	_
Animal pathways	10 (2)	6 (1)	_	15 (3)	_	_
Rocks/stones	5 (1)	12 (2)	_	50 (10)	24 (4)	_
Water	5 (1)	5 (1)	_	30 (6)	18 (3)	_
Tunnels	_	12 (2)	_	_	_	_
Treasure/valuable items	5 (1)	_	_	10 (2)	6 (1)	_
Other things left by people	5 (1)	_	_	5 (1)	6 (1)	_

Having made that point, it has to be conceded that in this instance, what might be considered to be a more sophisticated response on the part of the upper juniors - the inclusion of layers - is **not** attributable to intervention activities. The upper juniors were unanimous in this respect from the initial interviews.

Even accepting that the categories presented in Table 5.32 are derived from the responses of younger children, the degree to which the ideas are anthropocentric and domestically orientated is perhaps surprising. Young children, with their limited and circumscribed experiences, identify what is familiar to them. They project extensions of their home life, or parallels of that body of experience in the lives of animals. Living material and the artifacts of human and animal activity, sometimes romanticised or fantasised (perhaps succoured by the entertainment industry) are much in evidence. As for geology, substantial shifts in acknowledgement of the presence of rocks and stones and of water were recorded.

Table 5.33 Reference to Soil in 'No Layer' Drawings

	Pre-Intervention			Post-Intervention		
	Inf n=20	**LJ** n=17	**UJ** n=21	**Inf** n=20	**LJ** n=17	**UJ** n=21
Refers to soil but not by name	25 (5)	35 (6)	_	15 (3)	6 (1)	_
Specific mention of soil	30 (6)	_	_	50 (10)	12 (2)	_
Use of alternative name	_	12 (2)	_	10 (2)	18 (3)	_
'Layers' drawn	45 (9)	53 (9)	100 (21)	25 (5)	65 (11)	100 (21)

In their drawings, the younger children concentrated on the surface features, not venturing too far from home. It seems that when they considered what was below them in the ground, the young children considered, almost literally, what was beneath their feet. At least in part it seems legitimate to imagine that this more limited perspective is another reflection of the scale of 'world view' within which the younger children tend to operate. It is interesting to speculate on whether this scale could realistically be expected to be extended by teaching interventions. Teachers of young children are usually highly skilled in their sensitivities as to the limits of functioning within the age group. They **must** be in order to communicate effectively and set effective tasks. Other influences on children's ideas - and they are many and powerful, including books, comics, television and the cinema, indeed, whole sub-cultures of immense influence such as the Disney view of the world - reinforce, substantiate and extend children's limited perspectives **away from** the realities of the physical world. It is interesting to speculate about the extent to which it might be possible to extend young children's thinking in the direction of a more accurate understanding which is at the same time exciting, fascinating and makes contact with their imaginations.

5.3.3 Ideas about Depth

Understanding of a linear metric and ideas about distance and scale will constitute a limiting factor on children's conceptualisations of how far the Earth extends beneath their feet. Children's ideas about the distance the Earth extends below the crust are summarised in Table 5.34.

Table 5.34 Ideas About Distance Below the Earth's Crust

	Pre-Intervention			Post-Intervention		
	Inf n=20	**LJ** n=17	**UJ** n=21	**Inf** n=20	**LJ** n=17	**UJ** n=21
No quantification elicited but depth implied	30 (6)	18 (3)	10 (2)	5 (1)	18 (3)	10 (2)
Half cm - 1 Metre	5 (1)	6 (1)	_	_	_	_
1m - 10m	5 (1)	12 (2)	29 (6)	_	6 (1)	5 (1)
10m - 100m	10 (2)	35 (6)	24 (5)	10 (2)	6 (1)	5 (1)
100m - 1km	5 (1)	6 (1)	10 (2)	_	12 (2)	5 (1)
1km - 10km	_	12 (2)	_	_	_	_
10km - 100km	5 (1)	_	10 (2)	_	_	5 (1)
100km - 1000km	_	6 (1)	5 (1)	_	12 (2)	_
>1000km	_	6 (1)	14 (3)	_	6 (1)	10 (2)
No response concerning depth	40 (8)	_	_	85 (17)	41 (7)	62 (13)

It is immediately apparent that the order of magnitude within which those children who managed to offer any sort of quantified response were operating was far from appropriate. In thinking about the size of the Earth, there is little possibility of adopting the usual strategy of introducing children to measurement units, which is to start with meaningful but non-standard measures until such time as children have intuitions of the scale involved. That a few children were offering estimates in centimetres, metres or tens of metres confirms the difficulty children faced in engaging with this issue. Nor was there any indication of the situation being very different following the intervention activities. The activities to which teachers were drawn as intervention experiences tended to be those of a 'hands-on' nature; it might be an unrealistic challenge, in any event, to hope that young children's sense of geological scale might be sufficient to cope with the challenges posed in the time available. It does seem that this aspect of scale might be one which is developmentally constrained. Even so, some clues as to the provision of experiences which might enhance development would be welcome. It might be, for example, that distances between known places in the locality could be a starting point, extending to distances between known cities. Gradually, it might be possible for children to handle distances between countries or continents on the curved surface provided by a globe. Finally, such distances might be related to the size of the Earth itself, including radius, diameter and circumference. Until the metric is meaningful, there is little point in classroom transactions involving large distances. However, this is not to argue that the best policy is to wait for 'readiness', because children will be ready when they have built up the experiences and understanding in other areas. Challenges which **approach** the currently unattainable, but having the characteristic of being accessible, are needed.

It also seemed that the point at which many children stopped drawing was arbitrary, or, as in young children's typical portrayal of the sky, determined by the edge of the paper. This prompted interviewers to check whether what had been drawn was indeed intended to be the whole story. Only a minority, 16% pre-intervention and 9% after, and with no clear age association, were confident that what was drawn was all there was to be drawn as far as the depth of the subject was concerned. Perhaps as the result of their uncertainty, about one third of the sample offered suggestions about further materials existing beyond those which they had illustrated, when the question was posed.

No doubt the form individual children's drawings took was limited by the scope of their vision and imagery as well as their knowledge of what it was they wanted to represent. For example, the fact that none of the infants drew layers has already been discussed. As Table 5.35 shows, there were other differences in the formal elements of the drawing which are of interest. For example, none of the youngest children drew the centre of the Earth; their drawings retained the more parochial character referred to above.

Table 5.35 Nature of Drawings Which Depict the Centre of the Earth and Beyond

	Pre-Intervention			Post-Intervention		
	Inf n=20	LJ n=17	UJ n=21	Inf n=20	LJ n=17	UJ n=21
Show horizontal layers	–	12 (2)	43 (9)	–	29 (5)	19 (4)
Show concentric circles and/or curved layers	–	18 (3)	5 (1)	–	24 (4)	10 (2)
Show horizontal and curved layers	–	6 (1)	5 (1)	–	–	10 (2)
No drawings of this nature	100 (20)	64 (11)	47 (10)	100 (20)	47 (8)	61 (13)

Figure 5.2 Drawing Including Centre of the Earth

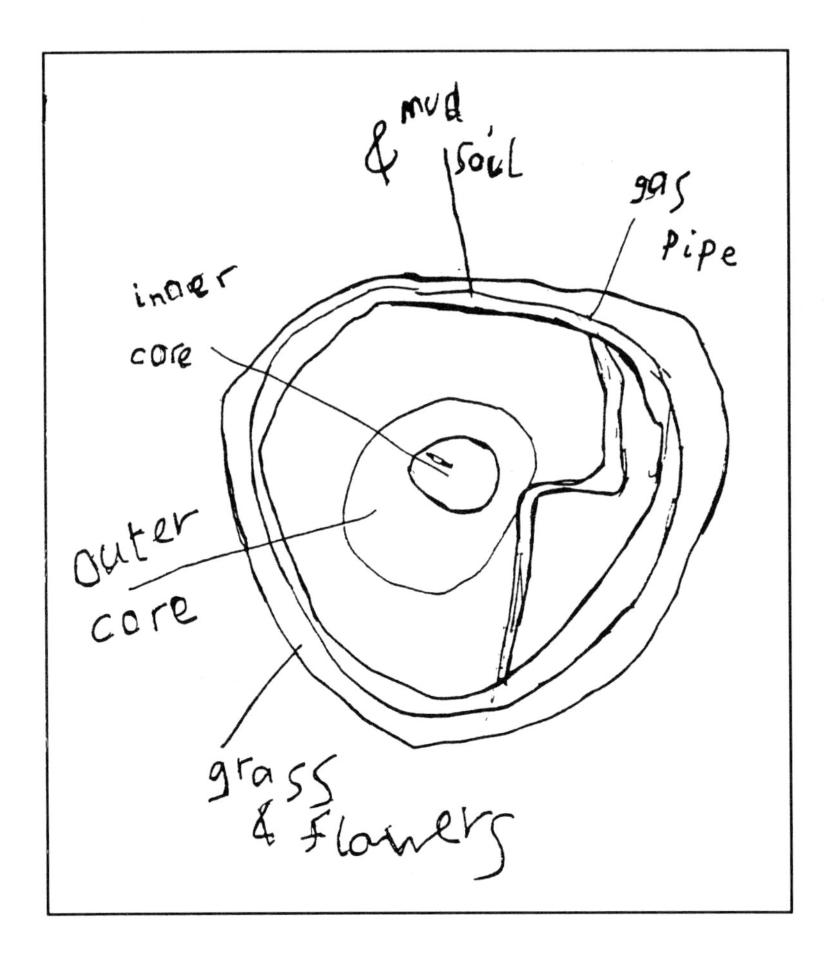

There are abundant images of the Earth, of the Moon, Sun and planets of the solar system available to children which show them to be at least circular, often convincingly spherical. Images depicting curved layers or cross section are likely to be less frequently encountered. Those children whose imagery supported such a representation without the support of secondary sources had accomplished something significant. (It is impossible to know what sources from inside or out of school children were tapping). The shifts resulting from the intervention were in fact very small and it probably remains to be seen what effects a concerted effort to offer representational support might have in this area of thinking.

5.3.4 Ideas about Temperature

Although no direct questions were posed which set a context of temperature or heat, a certain number of children introduced such ideas into the interview, as Table 5.36 indicates.

Table 5.36 Ideas About Heat in the Earth

	Pre-Intervention			Post-intervention		
	Inf n=20	LJ n=17	UJ n=21	Inf n=20	LJ n=17	UJ n=21
Heat mentioned in connection with lava/molten rock	_	18 (3)	19 (4)	5 (1)	24 (4)	5 (1)
Heat mentioned without reference to rock or lava	_	29 (5)	10 (2)	5 (1)	_	10 (2)
Rock (or lava) referred to as 'hot' (but not molten)	5 (1)	6 (1)	5 (1)	_	12 (2)	19 (4)
'Fire' referred to	_	_	10 (2)	_	_	5 (1)
Molten rock referred to without reference to heat	_	_	5 (1)	_	_	5 (1)
No reference to 'heat' or 'molten'	95 (19)	47 (8)	52 (11)	90 (18)	65 (11)	57 (12)

This was another area in which there was little evidence of change resulting from the intervention period. Indeed, there was a slightly reduced overall frequency of direct or indirect references to heat.

Table 5.36 also records the incidence of references to rock or lava in a molten state; these fluctuated from 14 to 12%. Children's ideas about the **location** of lava and/or molten rock are indicated in Table 5.37.

Table 5.37 Ideas About the Location of Lava/Molten Rock

	Pre-Intervention			Post-Intervention		
	Inf n=20	**LJ** n=17	**UJ** n=21	**Inf** n=20	**LJ** n=17	**UJ** n=21
In the core	_	24 (4)	10 (2)	_	12 (2)	10 (2)
Erupting through surface, on surface	5 (1)	_	5 (1)	_	_	_
In the core and mantle	_	6 (1)	_	_	12 (2)	5 (1)
Indeterminate below earth's surface	_	_	5 (1)	_	6 (1)	5 (1)
Other location	_	_	5 (1)	5 (1)	6 (1)	5 (1)
No reference to location of lava/molten rock	95 (19)	70 (12)	76 (17)	95 (19)	65 (11)	76 (16)

The locations which children indicated as sites of lava or molten rock included the core, the mantle, indeterminate locations below the Earth's surface and erupting through the surface in the form of volcanic activity. Only in the latter instance is there any chance of children gaining access to the kind of information which is likely to make immediate sense to them. Volcanic activity is spectacular and still and moving images of lava lakes, lava flow and volcanic eruption are striking and fairly familiar.

6. SUMMARY

6.0 Introduction

This report presents the outcomes of a series of collaborative research activities with teachers in the area of Earth Sciences, with a particular focus on children's ideas about rocks, soil and weather. The work on rocks and soil conducted in classrooms was supplemented by in-depth individual interviews; there was also an attempt to track any changes in children's ideas following an intervention period. The work on weather did not extend far beyond an initial exploration of currently prevailing ideas. It is emphasised that the strategies formulated in an attempt to help children to develop their ideas were an immediate response and as such, exploratory and tentative. The value of the research resides in the possibilities for further action which it may promote.

This chapter summarises the outcomes of the research together with brief commentaries and reflections.

6.1 Ideas about Soil

6.1.1 Properties and Composition of Soil

When presented with a sample of soil, half the youngest group responded by re-naming the material - 'dirt', mud', 'sand', 'dust', 'earth', etc. were common - this kind of response decreased with age. Conversely, more analytical responses relating to **origins**, **properties** or **composition** increased with age, more than half the upper juniors offered this type of response.

The other major response category associated with the infants was a description of the **function** of soil - most commonly concerned with the growing of plants.

Just over one third of the upper juniors described soil as a mixture. Nearly eighty per cent of the sample expressed the idea of the soil having changed in some way, referring to origins, change of location or some form of transformation. About one third of the sample cited changes of a superficial nature, alluding to colour, wetness, texture, and so on. Changes of geological relevance were cited with a much lower frequency, but it was useful to detect that such ideas are not totally alien to the age group. The ideas expressed included the following: internal movement of soil by creatures, enrichment by worms, 'clumping' of soil and rock formation, removal by wind and water.

Most children had nothing to say about the origins of soil, because it was not an issue for them. About a quarter of the upper juniors expressed an idea about soil having origins in another state, mostly inorganic in nature. The problems young children have in grappling with geological time scales is not surprising. A lack of awareness that soil formation is a constant process is perhaps another aspect of the same confusion with the time scales involved.

6.1.2 Changes in Children's Ideas about Soil

There was a shift in the perception of soil, from that of a homogeneous material to a perception of a complex mixture; the mean number of constituents identified pre-intervention was 2.6 while post-intervention this rose to 3.6. There was 'un-learning' as well as an increased appreciation of complexity.

For many children, first associations of soil were with its **function** in relation to growing plants; living plant material as a constituent of soil was suggested by about 50% of the sample overall. However, post-intervention, mention of living plant material as a soil constituent doubled amongst the lower juniors and halved amongst the upper juniors. It may be that older children were beginning to show a closer awareness of **components** of soil while the younger pupils were influenced by the **contiguity** of living plant material with soil as a growing medium. Fewer children offered living plant material as their **only** suggested constituent.

There was a large increase in the number of references to the living creatures to be found in soil. It seems likely that this awareness was the result of direct observation and examination of soil *in situ.*

Mention of inorganic materials in soil - rock in the form of sand, gravel, etc. - decreased by 20% amongst the infants, increased by an approximately similar proportion amongst the lower juniors and remained stable in the upper junior sample. References to human detritus declined from 9% to 5% overall. The overall mean number of suggestions of naturally occurring inorganic materials was at the level of 1.2 suggestions per interviewee. There was an impression of a shift away from specification of 'stones' to the more generic term, 'rock', though many children used both terms.

For many children, the perception was of soil as being a fine, brown, material variously referred to as 'mud', 'soil', 'earth', or 'dirt'. This was regarded as the **real** 'soil within the soil', the other material - stones and organic material - being treated as adulterations. The incidence of expression of this viewpoint fell from 33% to 21%, this shift being consistent with the possible adoption of more complex views of soil.

Another important shift, albeit within a minority of the total sample, was an increase from 10% pre-intervention to 25% showing recognition of the fact that soil contains different sized particles of the same inorganic constituent. This is regarded as an important understanding because it opens the possibility of a perception of the genesis of soil from larger to progressively smaller fragments of rock. Such a beginning must be complemented by ideas of change of location, mechanisms of fragmentation or erosion and the time scales involved.

The recognition that soil has a developmental history increased from 14% to 47%. All age groups showed gains, but particularly the lower juniors. The idea of **translocation** actually decreased in frequency, though this need not be interpreted as entirely retrograde, since the original ideas expressed about translocation tended to be anthropocentric. Since soil is so closely associated with people growing plants, its origins tended to be identified with centres of agriculture and horticulture - farms, fields, gardens and 'garden centres'. Such responses fell from 26% to 7% post-intervention. The diminution in assumptions about the role of human agencies in translocation of soil was even greater, but without a corresponding increase in appreciation of the part played by natural agents.

Some increased appreciation of the agency or mechanism of soil **transformation** was apparent. References to inorganic origins increased from 12% to 26%. There was a considerable increase in references to the effects of water, particularly the sea, and an increase in references to 'collision' of rocks. In view of the scale of energy transfers and geological scales time involved, it is not altogether surprising to find that those agents and mechanisms of transformation which were identified were those which are strikingly immediate in their power and effects. The mechanisms of transformation of rock can be identified as an important conceptual area in need of further consideration if children are to make progress in their understanding of soils.

Another approach to unravelling children's conceptions about the attributes of soil was through the presentation of samples in small phials of *sandy topsoil, sand from a sandpit, small pebbles, chalky soil* and *damp peat.* The inorganic constituents seemed to cause children doubts, post-intervention, possibly because they were looking for **mixtures**. For example, 17% accepted the sand as an example of soil pre-intervention, but none did so afterwards. The sandy loam and the presence of pebbles seemed to introduce a similar uncertainty. On the other hand, there was an increase in preparedness to accept peat as an instance of soil. Reliance on more superficial attributes - colour, for instance - as the criterial attribute in deciding whether or not to accept material as soil were less frequent, post-intervention.

6.2 Ideas about Rock

6.2.1 Properties of Rock

Children's first thoughts about rock related mostly to physical properties (45%) such as hardness and colour; location and origins were mentioned by less than 10%.

Complexities and ambiguities of language were responsible for at least some of the confusion experienced by children. Both 'rock' and 'stone' can refer **either** to single instances and fragments of material **or** the class of material. Informal vernacular conventions are in use which define the size and shape of an instance of a rock or a stone. When presented with a smooth piece of sandstone and a jagged piece of limestone, each about 200 ml in volume, a third of each of the younger groups were too uncertain to offer a response as to whether or not it was 'rock'; less than half the sample (all ages) accepted both specimens as being rock. The rounded sandstone sample was frequently rejected as an instance of rock on the basis of its smoothness and shape.

Children used bi-polar criteria in reaching their decisions as to whether or not a specimen was to be judged to be rock: rough/smooth, hard/soft, large/small, light/heavy, etc. These criteria were relatively superficial and were used inconsistently. They tended to be used inconsistently and as categorical rather than scalar attributes. The description of the attributes of rock specimens would be much more manageable if children were able to locate attributes on a **scale**. A more active form of observation (use of hand-lenses, files and balances) informed by more formal geological constructs was not in evidence. The data confirm that children lack a conceptual framework within which to consider and compare the attributes of rock.

6.2.2 Ideas about the Location and Extension of Rock Sources

Almost ninety per cent of the interview sample indicated a belief in the presence of rock underneath the place where the interview was taking place. However, it is probable that the majority of children responding in this manner did not have in mind an idea equivalent to that of 'bedrock'. Rather, their idea was of the presence of rock in the soil - a predominance of soil within which fragments of rock might be found. Some children were thinking of the building's foundations, with the words 'rock' and 'concrete' being used as synonyms. Probing suggested that even those who ventured the notion of a continuous layer of rock had a relatively localised idea in mind.

Most children's ideas about rocks and soil seem to be influenced by their local situation and daily experience. They lack a macroscopic notion of living on a planet largely comprised of rock, two thirds covered in water, the land masses having a thin scraping of soil in places, derived from local parent rock. Many children seem to hold the idea of living on land which is mostly soil, within which is to be found occasional masses of rock. There is even an occasional expression of the view that the very mountains sit on this soil. (Perhaps if our planet were known as 'Rock', rather than 'Earth', the situation might be a little different. More seriously, there are many interesting linguistic associations which might be explored cross-culturally).

No clear age-related pattern emerged with respect to the idea of continuous rock existing underground; 45% of the sample confirmed the idea while a further 15% were uncertain. However, probing revealed that children tended to find it difficult to envisage rock as ubiquitous; the scale seemed to be inconceivable.

6.2.3 Ideas about how Rocks Change

It was not surprising to find that younger children (40% of infants) found it difficult to articulate any scale or reference point in thinking about how long rocks have existed. About a quarter of the juniors interviewed suggested rocks have existed since the beginning of the Earth. There was only a very low incidence of processes central to the idea of geological change being referred to; weathering by water and the Sun were mentioned - the simple physical properties. There was little awareness of the large scale agents of geological changes, though a quarter of the upper juniors showed awareness of a role played by water. Physical breaking apart of rocks was alluded to by almost three quarters of the juniors, aggregation by 13%.

6.2.4 Changes in Children's Ideas about Rocks

Initially, it seemed that children were inclined to treat stones, pebbles, sand and rock as separate entities rather than as materials sharing the attributes of a common parent material. Post-intervention, the majority of the infants rejected the proposition that sand was an instance of rock; slightly more than half the juniors accepted sand as an instance of rock. The importance of size to children as a criterial attribute in deciding whether a material was to be regarded as rock seemed to diminish during the course of the study.

When presented with a rounded sample (a sandstone cobble) and asked whether or not it was rock, there was an increase in the number of children who asserted that it was. A major distraction for those who asserted that the sample was not rock was that of **vocabulary** associated with the size and form of the sample, some children suggesting that the name 'pebble' or 'stone' was more appropriate. Many of those who accepted the specimen as an instance of rock seemed to be basing their opinions on relatively superficial characteristics - hardness, smoothness and surface characteristics. It seems likely that most children have limited experience of the materials of which the planet they inhabit is made.

6.3 Ideas about the Earth's Structure

Children's ideas about what might lie beneath the surface of the Earth were explored using drawings. There was no constraint or suggestion that a circular cross-section would be appropriate. In fact, only 2% included the centre of the Earth in their drawings pre-intervention, this increasing to 9% post-intervention, with none of the infants included in this group.

There was a marked increase in the number of children representing continuous **rock layers**, though these gains were clearly associated with the older children in the sample. Perhaps their teachers felt that the use of secondary sources was more appropriate for the older children; perhaps the older children were more able to make sense of and retain this kind of information. It is not suggested that these layers of continuous rock were conceptualised as **global** in scale. Probing suggested that they were thought of as more confined in extent.

All the drawings produced by the upper juniors, both before and after intervention, were constructed in the form of layers. The most common representation was of layers including fragments and particles interposed with layers of rock. A variation was identified in which layers of rock, layers of particles and **technically named layers** were included. This latter classification showed a large increase amongst the lower juniors, post-intervention, as did the mention of soil. None of the infants and only about 40% of the juniors used technical terms such as 'core' (most frequently cited), 'mantle' and 'crust'.

There was a decrease in references to human detritus, from 10% to 4% amongst the juniors. Although relatively small, this shift represents a shift from anthropocentric and romanticised notions of buried treasure, archaeological remains and other human activity underground. The mean number of layers posited showed a strong positive correlation with age.

All the 'no layer' drawings were associated with younger children; their content showed a strong tendency towards anthropocentrism and a domestic orientation. They indicated the presence of living creatures, animal and plant remains, animal pathways, pipes and tunnels. Although they were not represented in layers, the youngest age group in the sample also revealed substantial shifts in their acknowledgement of the presence of geological features including rocks, stones and water. Young children were far more likely to draw precisely what was under their feet, and little further; they made far more frequent and specific references to soil in their drawings.

Invitations to estimate depths for the various features which children included in their drawings revealed that most of those who managed to offer any kind of quantified response had little sense of the scale involved.

There were age differences in the formal elements of the drawings which children offered. For example, as indicated above, only the juniors drew layers. Some of these layers were horizontal; some layers were concentric and curved; others drew a mixture of horizontal and curved layers. There were few shifts in the nature of children's drawings, pre- and post-intervention and the sources of information which they were utilising are not known. What it is possible for these children to comprehend and represent with understanding is in need of further exploration.

References to heat showed little change as the result of intervention; it was more frequently referred to by the older children in the sample, but remained in evidence at a level of less than half the upper juniors.

6.4 The Relationship between Rocks and Soil

Children's ideas about the relationship between rocks and soil were explored in the post-intervention interviews by asking them directly about the possibility of rock being transformed into soil. About one third of the sample confirmed the view that rock **could** be transformed into soil, this idea being more prevalent amongst older children. The most commonly cited evidence for denying the possibility was that rock is 'hard' while soil is 'soft'. Children probably have in mind the experience that rocks can hurt while soil would cushion a fall, so they deny the soil having perhaps their criterial attribute for 'rock', i.e., being 'hard'. A minority did suggest that rock becomes sand rather than soil; even fewer mentioned the need for the addition of other (organic) material for the rock derivative to be considered as soil.

Those who accepted that rock could be transformed showed awareness of rock crumbling or breaking down. This acceptance of the possibility of rock being convertible to a component of soil requires an appreciation that rock is not stable and immutable but is variable in size, shape and form over time. Few children mentioned volcanic activity, tectonic effects or glaciation. In part, this could be because their teachers were likely to favour direct experiences rather than secondary sources of information.

6.5 Weathering

Weathering effects are readily encountered in daily life within everyday environments, including urban settings. The variables involved in weathering effects are multiple and interacting. Children have to try to make sense of **agents, mechanisms** and **time scales** which might play a part in any particular weathering outcome. There may be a temptation to interpret the passage of time alone as a causal agent, since time **appears** to have such a significant independent impact on organic materials. (Intuitively, children are unlikely to think of human growth as related to the processes of the physical world in which a range of energy transfers occur. The same volatile gas - oxygen - which is associated with oxidation processes such as the rusting of metal is seen as life-giving in relation to human growth and activity. It makes more immediate sense to think of people simply becoming 'older' **as the result of** the passage of time). Environmental awareness has perhaps tended to put human agency centre-stage with a consequence that children are more sensitive to the outcomes of human activities than of other natural agencies which effect environmental change. Natural mechanisms are often massive in energy and effect - hurricanes, floods, earthquakes, volcanic activity, for example. More commonly, the mechanisms are massive in scale but imperceptible within the time frame of human experience - erosion by wind or tide, precipitation, the flow of water and the effect of freezing and expansion. Children's understanding seems likely to progress through exposure to a whole range of experiences of diverse weathering phenomena and a gradual appreciation of the scale of time, matter and energy transfer which may be involved.

6.6 Ideas about the Weather

Activity in this area provides some pointers for further research and development activities.

The youngest children tended to express unidimensional views about the weather, as though it were unchanging for any given day. They were able to generate simple iconic records of weather conditions, referring mostly to Sun, rain, wind and clouds. Juniors were able to devise symbols with accompanying written descriptions and awareness of **changes** in brightness, cloud cover, etc., as well as providing more **detailed** observations. A few showed awareness of quantified temperature and scales of wind force.

In their interpretations of weather maps, all numbers tended to be interpreted as referring to temperature. Arrows were generally recognised as relating to wind, but few children made inferences about wind direction or wind speed. The small minority who were able to discuss air pressure tended to be misled into making direct associations between high and low pressures and high and low temperatures.

Reflecting on a televised weather report, many infants showed no awareness of the report being a **forecast** or **prediction**, treating it instead as a **description**. Younger children also tended to assume that direct experience informed the weather report, invoking the use of astronauts, helicopters, aeroplanes, binoculars and cameras as the basis for the information collection preceding the report. Older children made more frequent mention of instruments such as thermometers, weather vanes and satellites. The notion that predicting the weather is a matter of finding out the experiences of those countries which experience the weather before it is met locally was encountered fairly frequently. This is probably a useful transitional idea since it is partially correct. However, in extreme form, it is expressed as the idea that the weather is static, being experienced by countries in turn as the surface of the Earth rotates 'underneath' the weather. The idea of weather as globally changing patterns of energy transfer was not encountered, nor expected, in the target age group.

There is evidence that children were tending to adopt gender stereotypes detrimental to women in relation to the expertise connected with the collection and reporting of weather data. (See McGuigan, 1992).

In explaining weather changes, children often regarded clouds as instrumental. In relation to rain, clouds were often described as containers of water, thunder as an impact between clouds.

The impact of the weather on people was described by the younger children primarily in terms of the **clothes** which prevailing conditions dictate. There was an awareness of potential physical hazards, including sunburn, catching colds, etc. The juniors showed awareness of more global effects - famine as the result of drought, coastlines threatened as the result of global warming and rising sea levels. It would appear that the human impact of these effects impinges on young children's awareness.

6.7 Concluding Comments

Since the research reported in this volume was conducted, it is increasingly likely that some of the participating children have lost their untutored innocence within the subject domains explored. The National Curriculum for England and Wales will have had an impact on many; various curriculum packages have been developed which address this particular area. The Earth Science Teachers' Association has also been very active. It seems realistic to expect considerable expansion of children's knowledge and understanding within this general area.

6.7.1 Opportunities for Direct Experience and Investigation

Within the enquiry into children's ideas about soil, the evidence suggests that there is much that is accessible, given more rigorous observation and some simple techniques. For most teachers, Earth Science has been an unfamiliar domain. Nonetheless, this subject area offers some attractive possibilities for those who are striving to adopt an active, investigatory approach to science. Both at the microscopic and macroscopic level, there are exciting possibilities for engaging children in absorbing and informative activities. The evidence is that children need to observe more widely and in more detail. Hand lenses and microscopes can readily challenge the prevailing idea that soil is a homogeneous material. Observation of the landscape can help children to observe and understand better the manner in which soil is distributed and its composition in relation to local geographical and geological conditions.

The study of rock forms is approachable even in a city environment. Building materials commonly include Portland Stone, slate, granite, marble and sandstone; stone masons may be approachable. Excavations are not uncommon in urban areas. Further afield, there are possibilities of visits to mines, quarries and natural features. Specialist centres with a specific focus on geological issues are available for school visits.

There is a strong case for children to have a wider experience of common rock forms, together with associated investigatory activities. Initially, these investigations might be those which children invent, discover and initiate themselves. There is also a case for some of the conceptual frameworks which geologists use being made available to children, provided these are intellectually accessible, safe and practicable. For example, the more detailed nature of rock, whether it is crystalline in structure, is relatively easily observable even with a hand lens. If children are to develop a greater understanding of rock, its structure, origins and transformations, they need to be encouraged to develop a more detailed and analytical approach, supported, when appropriate experiences have been made accessible, with the useful technical vocabulary that permits salient features to be compared and discussed.

6.7.2 Conventional Vocabulary and Techniques

Geologists and mineralogists use a whole range of techniques in identifying rocks and minerals, including reference to colour, 'streak' (the colour of the mineral in powder form), lustre, transparency, hardness, specific gravity, etc. Some of the vocabulary and conventions - Mohs' ten-point scale of hardness, for example - might be found to be accessible and useful to children in the primary age range.

Some of children's ideas about rock and soil seem to demonstrate scope for some fairly rapid gains. Confusions exist at the level of nomenclature which might be readily resolved by adopting some of the conventions used by geologists. For example, Figure 6.1 summarises how particle sizes are formally named and classified by geologists. It is not suggested that such a list is suitable for all children - the dimensions referred to would not be understood by younger children. The point is that a fixed reference **does** exist; as such, it offers some form of guideline towards teachers might wish to set their sights.

Figure 6.1 Definitions of Rock Fragments by Size

Clay *- a detrital mineral particle of any composition having a diameter of less than 0.004 mm.*

Silt *- a detrital particle finer than very fine sand and coarser than clay, in the range of 0.004 to 0.062 mm.*

Sand *- a detrital particle having a diameter in the range 0.062 to 2mm*

Granule *- a rock fragment of diameter 2 to 4 mm, larger than a coarse grain of sand and smaller than a pebble.*

Pebble *- a rock fragment, generally rounded, with a diameter of 4 to 64 mm.*

Cobble *- a rock fragment, rounded or abraded, between 64 to 256 mm in diameter. It is larger than a pebble and smaller than a boulder.*

Boulder *- a separated rock mass larger than a cobble, having a diameter greater than 256 mm. It is rounded in form or shaped by abrasion. Boulders are the largest rock forms recognised by sedimentologists.*

Although the scale might not be accessible to many children, it does seem appropriate to use the same **vocabulary** as geologists, particularly when vernacular descriptions confound existing confusions.

Another way of supporting children's observations, to take them beyond the superficial in the direction which will form useful foundations for future learning, is to encourage awareness of aspects such as the **grain size** of rock specimens. A very simple classification is as follows:

Grain Size

Coarse:	*over 2mm, (easy to see)*
Medium:	*0.2 - 2mm (visible to naked eye)*
Fine:	*Less than 0.2mm (visible only with a lens)*

It is not suggested that such a structure be **imposed** on children, but that the awareness of texture that many are likely to demonstrate be reinforced by teachers in this direction, in the knowledge that it makes geological sense to do so. (Igneous rocks are classified according to grain size and chemistry. For example, granite and gabbro are both coarse grained because they cool slowly but they contain different minerals which make granite acidic and gabbro basic. Gabbro and basalt are made of the same minerals but are classified as different rocks because basalt is fine grained as a result of its rapid cooling).

If children are to learn about the ways in which rocks are classified (igneous, sedimentary and metamorphic), their microscopic observations (detailed characteristics of rocks) and macroscopic observations (rock forms and features observable in the landscape) must be nurtured and encouraged, so that they can construct their understanding in appropriate directions.

6.7.3 Information from Secondary Sources

There is an important issue to be addressed as to how teachers might treat important geological processes - glaciation, for example - when these are not amenable to direct experience. It may be more fruitful to consider such processes case by case, rather than attempt generalisations. However, it is dubious that the orthodoxy that what cannot be directly experienced cannot be taught to children in the primary phase of education can be sustained if some aspects of Earth Science are to be approached honestly and effectively. The way forward is to explore secondary sources - models, illustrations, metaphors, computer graphics, etc. - imaginatively, to see what media may offer children accessible and comprehensible starting points. This is not to abandon constructivism, as children will make their own sense of secondary sources, just as they will of direct experience. It is a matter of searching for and exploiting effective **starting points** which children can challenge, discuss and make their own, so that their understanding is permitted to grow in useful directions - directions which underpin the development of a wider comprehension of Earth Science.

In time, children's imaginations may be able to encompass ideas about the origins of the solar system in general and the Earth in particular. Such understanding is very obviously **not** supported by the kind of experiential learning supported by most teachers of primary children. While the case for children's learning being supported by direct experiences is well established, it would be false to conclude that no other kind of learning is possible. Children clearly learn from sources other than direct experience: they learn from listening, from adopting the ideas implicit in language, from drama and literature, from reading and from electronic modes of information transmission. The constructivist view is that **whatever** the mode of information transfer, children make their own sense of their experiences. (Direct experiences which feel authentic can be misleading; more accurate interpretations of events are often counter-intuitive. The Earth's moon is perceived as a **source** of light by children, but is only a reflector; its 'seas' have never held water). Geological concepts call for ideas which are global in scale and which therefore cannot always be directly experienced.

Secondary sources of information are needed to offer children the conceptual frameworks to support the development of their own ideas. Such sources are not simple solutions; children make their own connections, as was evident in those who had confounded the Earth's core with more familiar apple cores. (There is an analogy which is sometimes used which compares the scale of the Earth's crust to the planet as a whole with the thickness of a skin on an apple; how far to pursue such analogies is often a matter of uncertainty!).

Our daily experiences and concerns tend to be parochial. With the advent of lunar landings and space probes, images of the Earth as seen from space are likely to be encountered by most children. However, even these images incline towards an impression of sea and cloud rather than the major solid mass of which the planet is composed. To understand the geological make-up of the planet, children need to have some form of representation of a cross section of the Earth.

Children showed awareness of some features of the structure of the Earth as represented in Figure 6.2. Teachers reported a lack of materials to meet children's appetite for information. As discussed in Chapter Three, children develop map making and map reading skills by starting from the domestic and parochial, extending their understanding as their experience and competence with topography develops. Is it realistic to think of a parallel development with respect to the structure of the Earth? This remains to be seen, for it is at least in part an empirical question. Thought needs to be given as to the important elements in Figure 6.2 which will serve to support understanding of important concepts; also, the manner in which such information might be made available. There are exciting possibilities offered by interactive computer graphics and virtual reality which could mirror available technology of space probes, satellite information and deep mineral exploration probes into the Earth's crust. This would be more inviting than Figure 6.2, as well as more exciting than the maudlin cartoon representations which currently dominate some children's thinking about what lies below the surface of the Earth.

Figure 6.2 Internal Structure of the Earth

The diagram below represents the patterns of differing density or elastic property of material as revealed by the measured passage of natural or artificial seismic impulses such as earthquakes or explosions.

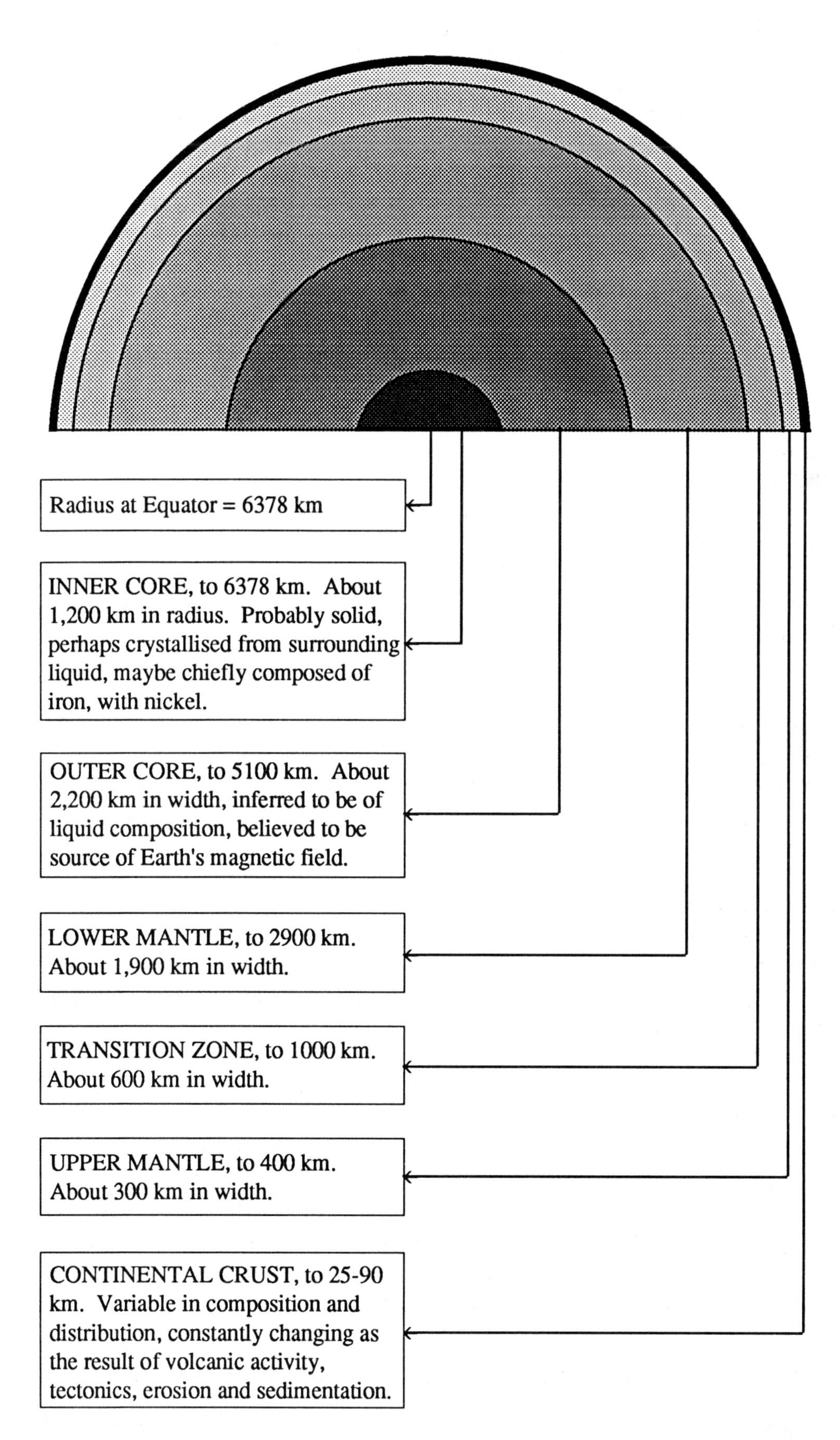

With regard to children's ideas about the weather, there is much that is familiar to children to build upon. There seem to be indications that they can move fairly readily from iconic to symbolic representations of weather phenomena. There seems to be scope for the development of quantified observations, growing out of more descriptive scales. The Beaufort scale provides an excellent ready-made example of an intuitively accessible scale which paves the way for a quantified approach to wind speed when children can manage it.

There are real difficulties for children in understanding weather phenomena which take account of air pressure. This domain is worthy of research in its own right. The invisibility and intangibility of air, the difficulties children have in comprehending that air has mass and exerts pressure; that the pressure varies and is associated with movements of 'fronts' on a global scale - each **component** presents problems. To understand and integrate all relevant elements is a formidable mental modelling task.

Secondary schools have already begun to explore the use of satellite data and telemetrics. Some schools have electronic mail links which are used in association with weather forecasting. Information technology would seem to offer much to complement direct experience.

List of Appendices

Rocks and Soil Research

APPENDIX I

SPACE SCHOOL PERSONNEL: Rocks and Soil

(for research carried out in 1990)

Lancashire LEA

County Advisor for Science: Mr P Garner

Advisory Teacher: Mrs R Morton

School	Head Teacher	Teachers
Brindle Gregson Lane Primary	Mr P Maddison	Mrs O Beal
Frenchwood Primary	Miss E Cowell	Mrs L Bibby Mrs C Pickering
Longton Junior	Mr M P Dickinson	Miss R Hamm Mrs L Whitby
Scarisbrick County	Mrs S Harrison	Mrs S Harrison Mrs A Stock
St Mary and St Benedict RC Primary	Mr M Sugden	Mrs F Flowers Mr A Hoyle
Walton-le-Dale County Primary	Mr A D Roberts	Mrs M Truscott
Mawdesley CE (Aided) Primary	Mrs M Ellams	Mrs M Ellams

APPENDIX II

SPACE SCHOOL PERSONNEL: Weather

(for research carried out in 1989)

Knowsley LEA

Inspector for Science:	Mr A Boyle
Advisory Teachers:	Mrs J Boden Mrs K Hartley

School	**Head Teacher**	**Teachers**
Brookside Primary	Mrs C Redmond	Mrs K Ford Ms M Smith
Holy Angels R C Primary	Mr J E Yoxall	Mrs M Fernley Mrs D Standley
Park View Junior	Mr M J Adams	Mr B Dixon Mrs D Fletcher Mrs D Pope
St John Fisher	Ms Ledsham	Mrs A Donnelly Mrs S King Mrs D Martin
St Michael's C E Junior	Mr D J Vernon	Mrs V Chadwick Mr M Tilling
Westvale County Primary	Mr J Barton	Mr M Jorgensen

Liverpool LEA

Matthew Arnold Infants	Mrs P Warbrick	Mrs M Ellams

APPENDIX III

EXPLORATION EXPERIENCES

ROCKS AND SOIL

---

1 Soil

A. Let children bring in samples of soil from their own gardens. NB there are safety considerations here - a need to be careful about the source of the soil and to wash hands after being near it etc. Let them display their samples at school. You may wish to add other samples for variety e.g. clay or sandy soils but do not show extremes (e.g. pure sand) because we may use these at the interview. Children could use magnifiers to look at the various samples. Let them describe and compare the various soils - they could draw pictures of a couple of samples and say/write a little bit about each one.

Product

Picture and description of a couple of soil samples.

B. Use two different looking soil samples from the display. Ask children how they would find out which one is better for plants to grow in. i.e. Let them think how they would do it rather than actually doing it.

Product

For each child, a written account of any investigation they can devise or for younger children, any comments they make on what they would do (or the answer they give if they already have one without an investigation!)

---

2 What is under where we are?

Take children out onto the school yard and ask them to imagine digging down through it - what would they find as they go down? Encourage them to go down as far as they can. Let them show what they think by drawings with annotations to make it clear. It would also be useful if you could talk to children about what they have drawn and where they have shown, say, tarmac or soil to get an estimate of how far down it goes, adding this to the drawing.
Repeat this exercise with the children standing on a patch of grass.

Product

For each child, annotated drawings of:

a. What is under the yard.

b. What is under a patch of grass.

3 Rocks

Ask children, **"where do we find rocks and what do they look like?"** Older children might write a few sentences about rocks (use the plural at this stage). Younger children might be asked to each speak out their ideas as part of a small group.

After this, ask each child if they could bring in a rock to display.

Product

a. Answers to the question **"where do we find rocks and what do they look like?"** written by children or noted by the teacher.

b. Notes on children's reactions to the rock bringing in task and the kinds of things they do bring in.

---

4 Weathering in the environment

Take children to look at the effects of the weather (without mentioning what has caused those effects!)

e.g. damaged plants/trees
church masonry and statuary
tombstones - engraved dates
brickwork
paintwork
railings
wooden structures

Let them draw pictures on one piece of paper of what they have seen and against each one give an answer to the question phrased to see if they can give a cause for what they see e.g **"Has it always looked like this? What was it like before? What made it change?"** For younger children, you might get everyone to draw the effects and ask for any causes from the subsample only.

Product

Pictures of weathered items with any explanation of what caused the change that the child can give.

APPENDIX IV

PRE-INTERVENTION INTERVIEW SCHEDULE

ROCKS AND SOIL

Aim: To ascertain children's views on the nature and origins of soil.

Show the child a sample of soil on a piece of paper.

1. **"What do you think soil is?"**

2. **"What do you think is in it?"**

3. Try to find out whether the child has any ideas about where soil comes from and whether it might always have been there by asking,

   (i) **"Has the soil always been there on the piece of ground where it was found"?**
   (ii) **"How do you think soil got to be like this?"**
   (iii) **"Where do you think soil comes from in the first place?**

4. Probe children's ideas about the permanence of soil. **"If we left this soil where it was found, do you think it would change at all?"**

Aim: To identify the range of materials accepted by children as soil.

5. For the samples A - E, ask **"which of these would you call soil?"** For these selected ask **"what makes you call these ones soil?"** and for those not selected **"what makes you say that this isn't soil?"**

Aim: To clarify the ideas in pictures children have drawn of what is under their feet.

6. Ask children to talk about their pictures showing what is under where we are. They may have drawn one or two pictures.

Try to:

(i) Get an idea of the scale of their drawing in terms of depths of any layers shown - **"How far down does that go?"**
(ii) Find the child's meaning for any terms used such as 'core'.
(iii) Clarify the contents of any layer if that is not already clear.
(iv) Probe as to whether they have gone down as far as possible by asking **"and is there anything under that etc?"**

Aim: To find out the child's understanding of the term 'rock' and to see what ideas they have about their origins and permanence.

7. Ask: **"What do you think of when I say the work 'rock'?"**

8. Show the jagged rock sample and ask **"Would you call this rock ?"** Probe any reasons given. If it's not accepted as rock also ask **"Is it a piece of rock, what do you think?"**

9. Repeat 8 for a smooth piece of rock.

10. Probe to see where children think rocks can be found. Ask **"Where can we find rocks?"** and see if they refer to 'special' places - then ask, **"Can we find rock under where we are now?".**

11. Ask:

(i) **How do you think rocks got there?**
(ii) **How long have they been here?**
(iii) **Do you think they will change at all?**

APPENDIX V

EXPLORATION EXPERIENCES

WEATHER

---

1 Record of Weather

A.Get each child to prepare a single sheet of paper divided into five; each section headed with the day of the week.

Ask children to keep a record of the weather for each day, using symbols, rather than words or pictures. You will probably want to use a synonym for symbols - feel free to explain what is required in a way which is appropriate for your class. Each child should, as far as possible, be allowed to decide how, what and when to record.

Product

One sheet from each child, showing five-day weather record.

Teacher annotations on the products from at least the sub-sample.

B.Activity One can be used to alert children to a change in the weather, from sunny to rain (or from bright to overcast failing a more dramatic change.) Ask children to use a comic strip format to draw an explanation of **how/why** the weather changes.

Children who have not used this drawing technique to express their ideas will benefit from an introduction from the teacher, that is, that the request is for ideas rather than pretty pictures; that the drawings can have labels or explanatory phrases; that adjacent frames indicate passage of time.

Product

One page comic strip type drawing showing change from sunny to wet weather.
Teacher annotations on at least the products of the sub-sample.

---

2 Television Weather Broadcast

Select a BBC1 Weather Summary (several occur during the morning and early afternoon). If it helps, use a video-recorder. Put the following questions to the children as a class.

1. **What was s/he doing? (that is, the broadcaster).**
2. **How does s/he know what the weather is like?**
3. **How does s/he know what the weather is going to be like tomorrow?**
4. **What do you think the various symbols mean?** (Express this idea in a way which your children will understand).

Teachers who feel that their children are capable of making a written response can present this as a written task, giving every help with expression/spelling, etc.

We will arrange a visit to infant classes to help with the collection of this information in other than written form.

Product

One response from each child answering the 4 questions.

Teacher annotations on at least the products of the sub-sample.

---

3 Creative Writing About the Effects of the Weather

Ask children to imagine that they are able to control the weather. They should be encouraged to explore imaginatively the short-term and long-term **effects** of their chosen weather pattern.

Drawings can also be encouraged, and the balance for younger children will be towards drawing rather than writing.

Product

A sheet of paper on which is some writing and/or drawing illustrating the **effects** of weather.

Teacher annotations.

APPENDIX VI

INTERVENTION GUIDELINES

ROCKS AND SOIL

---

1 Soil - a closer look

Aim: To encourage children to reconsider their ideas about what soil is, what it contains and how it is formed.

Using a sample of soil, children in groups might be asked to separate it and to name/classify the 'found' bits. They could then be asked to trace those components of soil back to their origins, prompting them with he questions "how did those 'bits' get in there?" Further activities could be tried to help them in thinking about classifying and tracking back.

Examples:

(i) Separating the soil into bits of different sizes using sieve/meshes/fabrics.

(ii) Adding water to look for floating components (the humus-organically formed part of soil).

(iii) Examining a compost heap, looking at various layers in it and considering what is happening. (formation of inorganic part of soil). Beware children thinking that soil is formed by people rubbing stones together.

(iv) Rubbing stones to see what happens (formation of inorganic part of soil). Beware children thinking that soil is formed by people rubbing stones together.

(v) Looking for small creatures in soil and thinking about what effect they have.

(If there was more time, further activities could show the presence of air (by adding soil to water), and water (by warming soil)).

Products:

1. Teacher's record of what was done.

2. Where possible, children might draw diagrams showing the various bits they found in soil and how they thought the various bits got in there (a sort of flow chart).

---

2 Soil - Which soil is better for plants to grow in?

Aim: To encourage children to develop their ability to fair test.

Children could plan once again an investigation to show which of two soils is better for plants to grow in, but this time working in groups. They should be prompted by the teacher's questions to be as rigorous as they can in their investigation.

- **How will you know which plant is growing better?**

- **What if your seed doesn't grow?** (if they choose only one seed).

- **How will you make your comparison fair?**

After planning, children should carry out the investigation. Again they should be encouraged to consider carefully what they are doing and to review their plans critically as they go along.

Products

1. Teacher record.

2. Agreed plans from each group of children + an account of the investigation itself (a group record.)

---

3 Rocks

Aim: To encourage children to think of rock as a material and to appreciate the variety of that material.

Most children take rock/rocks to be a piece/pieces of rock i.e. an isolated lump. Try to develop the notion of rock as a material by:

(i) Making a collage of pictures/photos of things made from rocks.
e.g. Statues, fire places, gravestones, doorsteps, kerbstones, pavements (some), lintels, buildings, roofing slates.

(ii) Classifying a set of materials into, for example, metal, plastic, rock.

(iii) Comparing different rock samples (as wide a variety of possible.)

(iv) Looking at pictures of rock strata (to help in idea of continuous bands of rock - see also activity 4.)

(v) Thinking about how rocks might have been formed (older children).

Try whichever of these seems suitable and whatever you can manage in the time available.

Product

Teacher record of the work done.

---

4 What's under the ground?

Aim: To encourage children to consider the structure of the Earth including the soil layer.

For some children the question "**What's under the ground?**" takes them into soil only. Others are able to go deeper down to consider the whole structure of the **Earth**. Without taking them down beyond their comprehension (!), try to get them to think about what is under the ground.

Experiences that could help are studies of:

People digging holes (roadworks etc.)
quarries
exposed rock in cliffs or hilly areas (memories of field trips)
mines and drilling
caves
channel tunnel (news)

Where first-hand experience cannot be given, then use secondary sources (books, photos, newspapers, videos e.g. of mining). One thing that appears difficult for children to appreciate is the massive layer of rock under the soil and the variation in soil depth which means that these layers are sometimes exposed and sometimes under a deep layer of soil.

Product

1. Teacher record of the work done.

APPENDIX VII

POST-INTERVENTION INTERVIEW SCHEDULE

ROCKS AND SOIL

Aim: To find out children's views of what is under the ground.

1 Ask the child to **draw a picture showing what is under the school yard**. Let them imagine that they are digging down and can go down as far as possible.

As they draw get them to clarify aspects of their drawing (eg, depth and terms used) and for the deepest thing drawn ask 'and is there anything under that?' etc.

Aim: To ascertain children's ideas on the nature and origins of soil.

2 Ask the children **to tell you about what they have been doing with soil and what they have found out about it.**

What do you think that tells you about soil?

3 Ask "**What do you think is in soil?**"

(Do this without a sample at first but if the child cannot reply, then prompt with a soil sample.)

4 Ask "**Where do you think soil comes from in the first place?**" to see whether the child has any ideas about the origins of soil.

5 Use samples of soil (sandy soil, chalky soil, peat) and ask "**Which of these would you call soil?**".

For those accepted, ask "**What makes you call these soil?**".

For those rejected, ask "**and what makes you say that this isn't soil?**".

Aim: To see if children are aware of ways to compare soils in a fair test.

6 **"How would you find out which soil is better for growing plants in?"**

Aim: To find out children's understanding of the term 'rock' and to see their ideas about origins of rocks.

7 Ask "**Would you call sand on the beach 'rock'?**". Ask for reasons for their answers - "**What makes you say that?**". Try to find their concept of rock - if it doesn't come up in answer, then ask "**What is rock like?**".

8 Show children the smooth piece of rock (pebble) - "**Would you call this rock?**". Probe any reasons given.

9 Ask "**How do you think rocks got here?**".

10 Ask "**Do you think rock can change into soil?**". "**What makes you say that?**".

11 Ask "**Where can we find rocks?**".

"Do you think there could be a rock stretching all the way underground from X (name and place) to Y (somewhere far away like Blackpool)?"

"What makes you say that?"

APPENDIX VIII

INTERVENTION GUIDELINES

WEATHER

We have to accept that weather is not a topic that lends itself readily to our usual form of intervention in which children test their hypotheses through careful control of variables. We propose therefore that your children record specific aspects of the weather and use the resulting data as starting points for further discussion of their ideas.

Activities

i) Each class constructs a form of "weather station". We suggest that the activity is introduced through a class discussion from which your children agree on the aspects of the weather that they are going to record at their station. Then the class divides into groups, each one accepting responsibility for recording one aspect.

ii) Each group needs to determine precisely what it is going to record and how it is going to carry out the measuring. Some groups will perhaps need to spend time designing and constructing measuring instruments though commercially produced measurers could be allowed. Groups may also conduct investigations/answer questions associated with their particular aspect of weather.

iii) Each group reports to the rest of the class on their measuring activity and their results.

iv) Each class produces a composite record of the results of all groups.

v) Each class uses the composite record as a basis for a discussion of linkages between various aspects of the weather and of causes for changes in the weather.

We predict that your children will suggest recording the following aspects of weather. Some suitable starting questions are also given for each one:

1 Sunshine (e.g. hours of sunshine/shadows/intensity)

- Is the sunshine always the same during the day?

2 Temperature

- How does the temperature change during the day?

3 Wind (e.g. speed/direction)

- Does the wind always come from the same direction?

4 Clouds (e.g. cover/shape/colour)

- What happens to the clouds during the day?

5 Rain (e.g. amount/drop size)

- When it rains is it always exactly the same?

Examples of questions for the final discussion

Is it always colder when the wind blows?

Does it always rain when it's cloudy?

Is it always hot when the sun shines?

What is the weather like when it's dark?

What do you think is going to happen to the clouds we can see today?

What has happened to the clouds we saw yesterday?

What happens to the sun at the end of the day?

Products

1. From each child; an account of what they understood they were measuring, how they measured it and any investigations they carried out. (Drawings, prose, annotated diagrams, tape recordings would all be acceptable.)

   Please annotate the work of your sub-sample, if necessary.

   ii) From each class: the composite chart of the groups' recordings.

   iii) From each teacher: notes on the major ideas emerging from the final class discussion and annotations on any work from the sub-sample.

INTERVENTION RECORD SHEET 1

INTERVENTION INVESTIGATIONS

Date:__________ School Number and Class:____________________

Names of Children involved:
(indicate numbers on class list)

General Question:

Specific Question to be Investigated:

Brief outline of Children's Plan:

Results of Children's Investigation:

Children's Interpretation of Results:

Any Further Investigations Suggested:

INTERVENTION RECORD SHEET 2

CLASS DISCUSSION SUMMARY SHEET

TEACHER.................... DATE...................

NO OF CHILDREN...... SCHOOL NO................................

QUESTION..

Main Ideas No supporting Ideas

INTERVENTION RECORD SHEET 3

VOCABULARY

Date:__________ School Number and Class:____________________

Names of Children involved:
(indicate numbers on class list)

Word to be Explored

Context in which word used

Definitions Suggested by Children:

1.

2.

3.

Activity:

Outcome of Activity in Terms of Children's Definitions:

APPENDIX IX

BIBLIOGRAPHY

Bar, V. (1989) Children's views about the water cycle. *Science Education* 73 (4) p481 - 500.

Baxter, J. (1989) Children's understanding of astronomical events. *International Journal of Science Education* 11, Special Issue, p502 -513.

Department of Education and Science and the Welsh Office (1989) *Science in the National Curriculum*, H.M.S.O., London.

Department of Education and Science and the Welsh Office (1991) *Science in the National Curriculum*, H.M.S.O., London.

Happs, J.C. (1981) Soils, *Science Education Research Unit Working paper 201*, University of Waikato, New Zealand.

Happs, J.C. (1982a) Mountains, *Science Education Research Unit Working paper 202*, University of Waikato, New Zealand.

Happs, J.C. (1982b) Rocks and minerals, *Science Education Research Unit Working paper 204*, Waikato University, New Zealand.

Happs, J.C. (1984) Soil genesis and development: views held by New Zealand students. *Journal of Geography* 83 (4), p177 - 180.

Happs, J.C. (1985) Regression on learning outcomes: some examples from the earth sciences, *European Journal Science Education 7* (4) p431 - 443.

Leather, A.D. (1987) Views of the nature and origin of earthquakes and oil held by eleven to seventeen year olds, *Geology Teaching 12* (3), p102 - 108.

McGuigan, L. (1991) Fair Weather Reporting, *Primary Science Review,* No 19.

Moyle, R. (1980) Weather, *Learning in Science Project Working paper 21*, University of Waikato, New Zealand.

Nelson, B.D. Aron, R.H. and Francek, M.A. (1992), Clarification of selected misconceptions in physical geography, *Journal of Geography 91* (2), p76 - 80.

Piaget, J. (1929) *The child's conception of the world,* Routledge and Kegan Paul Ltd, London.

Piaget, J. (1930) *The child's conception of physical causality*, Kegan Paul, Trench, Trubner and Co. Ltd, London.

Russell, T., Longden, K. and McGuigan, L. (1990), *Materials*, Primary SPACE Project Research Report, Liverpool University Press, Liverpool.

Russell T. and Watt D. (1990) *Evaporation and Condensation*, Primary SPACE Project Research Report, Liverpool University Press, Liverpool.

Stepans, J. and Kuehn, C. (1985) What research says: children's conceptions of weather, *Science and Children* 23 (1), p44 - 47.